漫食光——餐厅平面与空间设计

Romantic Dining Time

Restaurant's Graphic and Space Design

（意）托马斯·拉玛诺斯卡斯 编　常文心 译

辽宁科学技术出版社

MORE THAN FOOD, MORE THAN DESIGN

不仅是美食，不仅是设计

Restaurant is a peculiar place and it is never just about the food. Even no thrills trattoria in rural Italy is much more than just a spot to have a tasty meal. It's a social act, gesture of kindness, multi-sensory experience, conversation greaser, aesthetic pleasure.

To think of it, it is more of a spectacle (with variations in how lavish the production is) rather than a station of edible fuel for your work. Even when you eat you're not just eating.

Let's observe. You search for options or hear recommendations – "check out this amazing Polynesian cuisine place", "miso skewers are so in", "go to raw medieval fish pub". You peak into their website which hopefully doesn't use flash animation, check the photos on Facebook or Instagram, study the menu. You read the reviews by the critics or "friend critics", and you make your decision – this is the place I want to spend my hour or two in.

Then, it's the show time – your lunch or dinner starts. The signage greets you from a far, it assures you – the right choice was made. Somebody will meet you and they may or may not be dressed in suit, but they will mostly be polite and welcoming. You will be lead to your spectator's or rather, participator's seat. The show begins.

The table, decorated or plain in its own vintageness, the indifferent or super friendly personnel, the introduction of today's specials or absence of them, the little rituals of food, the order in which it arrives, your final cup of espresso and decisive "check, please" – these are just a few of touch-points you'll encounter during this performance. And it all can be designed in different ways – deliberate show-off, understated quality, or down-to-earth simplicity.

All your senses are participating in this show – you look, you smell, you taste, you touch. You're not here solely for the food. You came to eat, yes, but also talk, sit in awkward silence, mumble under your nose. Or: confess, propose, or chat for hours until the manager politely informs you that it's way past the closing hour. This is a social deed. You are not cooking, you are in the open, seen here, with your knife and fork, performing the eating.

What a strange yet fundamentally beautiful expression of gratitude this is – if you had a nice meal, you thank the chef or the staff, they thank you in return, some will be are genuinely happy. This simple exchange could influence how your day unfolds.

Great branding is a way for the restaurant to accentuate the things it wants to stress, to

餐厅是一个独特的场所，它不仅仅意味着美食。即使是意大利乡村平凡的小餐馆也绝不仅是一个享用美食的场所。就餐是一种社会行为，一种友好的姿态，是多重的感官体验，是社交的润滑剂，是美的享受。

餐厅更像是一个琳琅满目、丰富多彩的场面，而不是一个为工作添加食物原料的加油站。即使当你正在吃东西时，你所做的也不仅仅是"吃"这个动作。

让我们仔细观察。你会搜索选项或听取建议——"试试这家波利尼西亚餐馆""现在很流行味噌烤串""去中世纪风的海鲜酒馆吧"。你去网上搜索它们的网站，在社交网站上看照片，研究菜单，阅读它们的美食评价，然后做出选择——我愿意在这里度过一两个小时。

然后就是表演时间了，你的午餐或晚餐由此开幕。餐厅的招牌远远地就向你打招呼，它向你保证你的选择是对的。门口穿或不穿着套装的服务员以礼貌友好的姿态接待你。你被引到自己的座位，表演开始。

装饰精美或朴素复古的餐桌、冷漠或超级亲切的店员、每日精选的推荐、用餐礼仪、上菜顺序、最后的一杯意大利浓咖啡以及决定性的"买单"，这一切都只是整个表演中的部分节点，它们的设计方式多种多样——刻意的炫耀、低调高尚、简洁朴实。

你的所有感官都参与到了表演之中——你看，你嗅，你品尝，你触摸。你不仅是为了美食而来到这里。除了吃之外，你还会谈话，陷入尴尬的沉默，含糊地低语，忏悔，求婚，或是海阔天空地聊天，直到餐厅经理礼貌地通知你餐厅早已打烊。这是一种社交行为。你不是在烹饪，而是手拿刀叉，表演就餐。

就餐结束后，略显奇怪却又十分重要的环节就是表达感谢。如果你的就餐过程十分愉快，你会感谢主厨和服务员，他们也会感谢你，一些人会真心感到高兴。这种简单的交换可能会影响

underline the important, to orchestrate the overall feel and to balance-out the expectations. Besides the actual food, there are dozens of the ways we experience dinning. When creating this ensemble of tiny details, first of all we talk about the philosophy of restaurant's food.

From this we move onto the visitors (participators!) which the place wants to attract. The two elements dissolve into many others: what are our observable assets, how are we interacted with online and on mobile, how are we seen from the outside and inside, what pieces of information reach the table, in what order and so on. Apparently, this new restaurant is not the only one in particular area of the city, so it needs to stand out, right? And that's just the visual cues, words are of great importance too, the way restaurant talks is just as essential.

We go further: what is our name, how do we explicitly explain the ideology of the place in two sentences, how we present the items on the menu or special offers? The language should be the entourage of the visible.

The worst thing branding can do is to over-promise or miss the point, become too decorative when it needs straight-forwardness, or too primitive when subtlety was required, and to lose coherency and consistency through various pieces of application. And then you, the one who participates in this nourishing spectacle, are left confused. This is not what you expected. This is not good.

On the other hand, the ideal in branding is for it to be almost invisible. The less noticeable it is on its own, the better and more effective it becomes. This doesn't mean you have to be bland. This means you don't overshadow what is on the plate. I see branding as a creative enhancer affecting the taste buds of a customer. Everything feels right, you are content and, of course, your review will start with the words: "Only the food matters and here it's sensational". This is how we know we did our work as good as we can. We didn't get in the way of enjoying the food. And we're glad you will be back for more.

Tomas Ramanauskas
Co-founder, NEW! Creative Agency

你一天的心情。

优秀的品牌设计让餐厅得以突出特色，强调重点，协调整体感觉，平衡预期效果。除了实在的美食之外，我们还能用许多种方式体验就餐。在整合这些细节时，我们首先会谈到餐厅的美食哲学。

下一步我们将考虑到餐厅渴望吸引的来访者（即参与者）。这两个元素将融入许多其他的元素：我们的可见资产是什么？我们如何实现网络和移动网络互动？以怎样的顺序实现……显然，这家新餐厅不是城里唯一的特别场所，所以它需要脱颖而出。与视觉信号相同，文字的作用也至关重要，餐厅的表达方式是最基本的特色。

然后我们开始进一步的设计：餐厅用什么名字？如何用三言两语来说明餐厅的经营哲学？如何展示菜单内容和特色菜？语言的表达应当符合视觉环境。

失败的品牌设计要么过度许诺，要么不得要点，在需要直白的时候太过浮夸，在需要精致的时候过于朴实，丧失了各种设计应用之间的凝聚感和一致性。在这种繁杂的信息背景下，参与者会感到迷茫，认为这并不是你想要的，从而对餐厅失去信心。

另一方面，理想的品牌设计近乎隐形。品牌营销越低调，营销效果越有效。这并不意味着必须平淡无味，而是意味着品牌设计不能凌驾于菜品之上。我把品牌设计看成一种创新的强化剂，它能影响消费者的味蕾。当一切都感觉良好，消费者感到满意时，评论的开头当然是："美食才是重点，令人感动。"这样我们就知道自己的工作没有白做。我们不会妨碍你享用美食，而且期待你下次的到来。

托马斯·拉玛诺斯卡斯
NEW! 创意公司联合创始人

CONTENTS 目录

MEXICAN RESTAURANT 墨西哥餐厅

ALBERTO SENTIES CATERING 018
MONTERREY, MEXICO
Odd Foil Patch as Regional Symbol
阿尔贝托·桑提斯餐饮
墨西哥，蒙特雷
具有地域象征的奇怪银箔图形

EL CARINITO 022
SANTIAGO DE QUERÉTARO, MEXICO
Dynamic Mexican Handmade Graphics
亲爱的餐厅
墨西哥，克雷塔罗
具有活力墨西哥色彩的手绘图形

LA PEÑA DEL ROSARIO 026
PUEBLA, MEXICO
A Pure and Enigmatic Design
罗萨里奥餐厅
墨西哥，普埃布拉
返璞归真的设计质感

PUEBLA 109 028
MEXICO CITY, MEXICO
Classic Symbols from the Age of Mexican Philately
普埃布拉 109 号餐厅
墨西哥，墨西哥城
墨西哥集邮时代的经典符号

CALEXICO'S 032
STOCKHOLM, SWEDEN
A Mix of Street and Luxury
卡勒西可餐厅
瑞典，斯德哥尔摩
街头感爆棚的豪华餐厅

MEXOUT 034
SINGAPORE
A Mexican Restaurant with Eccentric Thoughts and Design
墨西哥人餐厅
新加坡
充满古怪思想与设计的墨西哥餐厅

LUCHA LOCO 038
SINGAPORE
Minimalism Provides an Original Feeling
疯狂摔跤餐厅
新加坡
极简主义赋予的“原生态”感觉

CASA VIRGINIA 040
MEXICO CITY, MEXICO
Detail Is a Sort of Language
弗吉尼亚私房菜馆
墨西哥，墨西哥城
细节也是一种语言

MONTAGU 042
MEXICO CITY, MEXICO
Design Inspiration from Travel Experience around the World!
蒙塔古餐厅
墨西哥，墨西哥城
设计灵感来源于主厨环游世界的经历！

CONTENTS

SPANISH RESTAURANT 西班牙餐厅

TAMARINDO 046
OURENSE, SPAIN
Adding Fresh and Soft Ice-cream Colours to a Restaurant with Rainy Galician Qualities

塔马林多餐厅
西班牙，奥伦塞
为仿若加利西亚阴雨连绵般气质的餐厅添加一抹清新柔和的冰淇淋色

VINO VERITAS 050
OSLO, NORWAY
Design Is as Natural as "Ripening of Grape"

维利塔斯葡萄酒餐厅
挪威，奥斯陆
设计应该是像"葡萄成熟了"一样自然的事

SAL CURIOSO 054
HONG KONG, CHINA
To Show Unique Culinary Experiments in Illustrations

萨尔·库里奥索餐厅
中国，香港
以插画形式呈现独特烹饪实验

FRENCH RESTAURANT 法式餐厅

GALO 058
MONTERREY, MEXICO
Homage from Zeppelin Icon

加洛小厨
墨西哥，蒙特雷
齐柏林式飞艇图标的敬意

THE WOODSPEEN 062
NEWBURY, UK
Michelin Star Chef John Campbell's New Venue: Simple Elegance

伍德斯宾餐厅
英国，纽伯里
米其林星级主厨的新餐厅：简洁的优雅

BALZAC 064
SINGAPORE
Humanist Consciousness of a French Restaurant

巴尔扎克餐厅
新加坡
一家法式餐厅的人文情怀

QUALINOS 068
ZURICH, SWIZERLAND
Inspiration from French Pot "Le Creuset"

加里诺斯餐厅
瑞士，苏黎世
法式"酷彩"锅带来的灵感

LOUIS 070
ZURICH, SWIZERLAND
Stylish Simplicity

路易斯啤酒屋
瑞士，苏黎世
有格调的简约

FRU FRU 072
FERMO, ITALY
A Combination of Traditional French Kitchen Values and Homemade Cooking

法鲁餐厅
意大利，费尔莫
传统法式烹饪与家常料理在设计中的意境融彻

BRITISH RESTAURANT 英伦餐厅

THE COLLECTION 076
LONDON, UK
Ultimate Fashion in Whiting

收藏餐厅
英国，伦敦
清水抹白的极致时尚

HAY MARKET 080
HONG KONG, CHINA
Vintage British Style Created by Geometric Shapes and Classic Letterforms

干草市场餐厅
中国，香港
几何图形与古典字体打造复古英伦风

ITALIAN RESTAURANT 意大利餐厅

SOPRA 086
SINGAPORE
Imaginations of Post-war Italy

索普拉餐厅
新加坡
战后意大利的影像

THE SICILIAN 088
New South Wales, Australia
The Gentlemen's Style of 1940's Gangster Films

西西里餐厅
澳大利亚，新南威尔士
20 世纪 40 年代黑帮电影的绅士风格

UN POCO 090
STOCKHOLM, SWEDEN
New York's Rawness VS Swedish Simpleness

一点点餐厅
瑞典，斯德哥尔摩
纽约的狂野碰撞瑞典的简约

IANNILLI 092
MONTERREY, MEXICO
"Romantic Nostalgia" Concept Created by Delicate Craftsmanship

伊安尼里餐厅
墨西哥，蒙特雷
精致工艺打造“浪漫怀旧”主题风格

GRASSA 094
PORTLAND, USA
Perceptual Feeling of Commercial Illustration

格拉萨餐厅
美国，波特兰
商业插画的直观情感

ESTE OESTE 096
LISBON, PORTUGAL
Design Expression of Cultural Variety

东西餐厅
葡萄牙，里斯本
多样性文化的设计体现

CONTENTS

JAPANESE RESTAURANT
日料店

50%,TRANSLUCENT RESTAURANT 100
TOKYO, JAPAN
A Fresh and Aesthetic Restaurant in Tokyo

50% 透明餐厅
日本，东京
"人间有味是清欢"，来自东京的清新餐厅

TANUKI RAW 104
SINGAPORE
The Tanuki and Raw Industrial Style

狸猫原生餐厅
新加坡
一只狸猫与"原生"工业风格

SUSHI & CO. 106
BALTIC SEA
A Combination of Scandinavian Elements and Oceanic Symbols

寿司公司
波罗的海
斯堪的纳维亚元素与海洋元素的交融

FAT COW 110
SINGAPORE
The Japanese Aesthetic – Wabi Sabi

肥牛餐厅
新加坡
日本美学的"残缺之美"

MIU CREATIVE CUISINE 112
BEIJING, CHINA
Wonderful "米" Shape Graphics

MIU 秘团
中国，北京
奇妙的米字图形

NAGOYA 114
PARIS, FRANCE
Natural Beauty of Oriental Zen

名古屋日本料理
法国，巴黎
东方禅意的自然之美

MATSUYAMA 116
PARIS, FRANCE
Chinese Ink Painting as An Exotic Symbol in the West

松山日本料理
法国，巴黎
水墨元素在西方的异域象征

NOZOMI SUSHI BAR 118
VALENCIA, SPAIN
"Emotional Classic" and "Rational Contemporary"

希望寿司
西班牙，巴伦西亚
"感性的古典"与"理性的现代"的双重性

TORO TORO 122
MONTERREY, MEXICO
Visual Environment Reminiscent of a Night in Tokyo

鲔鱼餐厅
墨西哥，蒙特雷
东京夜生活般的视觉环境

MEDITERRANEAN RESTAURANT 地中海餐厅

KESSALAO 126
BONN, GERMANY
A Colourful Restaurant from the City of Beethoven
凯斯萨劳快餐店
德国，波恩
来自贝多芬故乡具有糖果般色彩缤纷的餐厅

LA BICICLETA 130
CANTABRIA, SPAIN
The Beauty of Geometry
自行车餐厅
西班牙，坎塔布里亚
几何图形之美

SEAFOOD RESTAURANT 海鲜餐厅

COSTA NUEVA 134
NUEVO LEON, MEXICO
Mexico's Progressive and Modern Artistic Boom in 1950s
新海岸餐厅
墨西哥，新莱昂
墨西哥 20 世纪 50 年代激进而又现代的艺术浪潮

COSTA CHICA 138
NUEVO LEON, MEXICO
Most Intuitive Visual Language
小海岸餐厅
墨西哥，新莱昂
最直观的视觉语言

BARBA 140
DUBROVNIK, CROATIA
A Relaxed Atmosphere of Modern Croatian Design
巴尔巴餐厅
克罗地亚，杜布罗夫尼克
现代克罗地亚设计风格营造的轻松氛围

OXLOT 9 142
COVINGTON, USA
Etching Style Expresses Classic Nature
奥克斯洛特 9 号餐厅
克罗地亚，杜布罗夫尼克
蚀刻版画风格展现古典内涵

BISTRO 小酒馆

LA VACHE! 146
HONG KONG, CHINA
Inspired by French Cartoonist Jean-Jacques Sempé!
好好餐厅
中国，香港
灵感来自于法国漫画家让 - 雅克 · 桑贝！

THE RUSTIC BISTRO 148
SINGAPORE
Black and White in Deconstructivism and Post-modernism
乡村酒馆
新加坡
解构主义与后现代风格里的黑白画映

ENOTECA SAN MARCO 150
LAS VEGAS, USA
An Eclectic Mix of Typefaces
圣马尔科红酒餐厅
美国，拉斯维加斯
字体混合的奇妙设计

CONTENTS

ISLAMIC RESTAURANT 清真餐厅

HABIBIS 154
SAN PEDRO GARZA GARCIA, MEXICO
An Arabic-Mexican Fusion Taqueria Using Gentle Typeface to Show Friendliness
哈比比斯餐厅
墨西哥，圣佩德罗
柔和字体打造亲切感的清真餐厅

SERAI 158
KUALA LUMPUR, MALAYSIA
"Simple yet Incomparably Beautiful"
客栈餐厅
马来西亚，吉隆坡
"朴素而天下莫能与之争美"

VEGETARIAN RESTAURANT 素食餐厅

RESTAURANTE BAOBAB 162
MADRID, SPAIN
A Vegetarian Restaurant with Humanism
猴面包餐厅
西班牙，马德里
人本主义的素食餐厅

FAST-FOOD RESTAURANT 快餐店

SIMPLE 168
KIEV, UKRAINE
All Details Sustain and Complement Each Other
简单餐厅
乌克兰，基辅
极致细节下的相辅相成

DRAMA BURGER 172
OVERLAND PARK, USA
A Dramatic Burger Restaurant
戏剧汉堡
美国，欧弗兰帕克
戏剧化效果的独特汉堡店

WING'N IT NYC 176
NEW YORK, USA
Freestyle Typeface
纽约飞翔餐车
美国，纽约
自由"字"在

CHICK-A-BIDDY 178
ATLANTA, USA
Bright Illustrations for Fast Food
小鸡餐厅
美国，亚特兰大
明亮的快餐插画

EL CAMINO 182
SAN PEDRO GARZA GARCIA, MEXICO
Sustainable Graphic Language
埃尔·卡米诺餐车
墨西哥，圣佩德罗加尔萨加西亚
可持续图形语言

BYRON 184
LONDON, UK
Hand-drawn Type and Illustrations Catering for the Restaurant
拜伦餐厅
英国，伦敦
手绘文字和插画迎合餐厅氛围

SMACK LOBSTER ROLL 186
LONDON, UK
Inspired by Signage on Boats
斯迈克龙虾卷餐厅
英国，伦敦
灵感来自船体手绘标识

UNFORKED 188
KANSAS, USA
A Playful, Distinct Language to Create a High-quality Fast Food Brand
无刀叉餐厅
美国，堪萨斯
独特有趣的设计语言打造高品质快餐品牌

PIZZA HOUSE

比萨店

LE PARI'S 190
BRUSSELS, BELGIUM
Lovely Patterns and Fresh Style

帕里快餐厅
比利时，布鲁塞尔
可爱图案演绎清新风格

HOLLY BURGER 192
SAN SEBASTIÁN, SPAIN
Inspired by the Banana Leaf Wallpaper of the Beverly Hills Hotel in Los Angeles

奥利汉堡
西班牙，圣塞巴斯蒂安
灵感来自美国洛杉矶比弗利山酒店的香蕉叶墙纸

THE FITZGERALD BURGER CO. 194
VALENCIA, SPAIN
A Premium Vintage Touch by Hand-drawn Lettering with Brushpen

菲茨杰拉德汉堡餐厅
西班牙，巴伦西亚
毛笔手绘文字与图标插画所缔造的另类复古风格

EMBUTIQUE 196
LONDON, UK
Hipstery London Style

艾姆布提克餐厅
英国，伦敦
时髦的伦敦风格

THE BITE 198
ZURICH, SWITZERLAND
Strong Industrial Feeling in Black and White

一口餐厅
瑞士，苏黎世
黑白色彩下的浓重工业气息

CHOP CHOP 200
TEL AVIV, ISRAEL
Collage in Small Space

乔普餐厅
以色列，特拉维夫
拼贴画融入小空间

FISHCAKE 202
SEOUL, KOREA
Simplicity Is Ultimate

鱼饼店
韩国，首尔
简约，至上

KOODOO 206
MOSCOW, RUSSIA
Koodoo and His Team!

弯角羚餐厅
俄罗斯，莫斯科
弯角羚和它的团队们!

LA VITA IN FIORI 208
BARCELONA, SPAIN
Symbolic Christmas Colours

菲奥里生活餐厅
西班牙，巴塞罗那
标志性的圣诞色

NICKS 210
RIO CLARO, BRASIL
A Long-range Design for Mr. Nicolau

尼科斯餐厅
巴西，里奥克拉鲁
为尼古拉先生的长远目标做设计

PIZZA BARBONI 212
JAKARTA, INDONESIA
Vintage Style in Street Illustrations

巴博尼比萨
印度尼西亚，雅加达
街头插画的复古情怀

CONTENTS

BARBECUE RESTAURANT 烧烤店

LA FAMA BARBECUE 216
BOGOTA, COLOMBIA
Perfect Combination of Bogota Local Butchery and "Homely" Concept
肉店烤肉餐厅
哥伦比亚，波哥大
波哥大本土肉店风格与"居家理念"的完美混搭

STREETZ AMERICAN GRILL 218
HOPKINS, USA
A Fashion and Friendly Feeling Created by American Vintage Style
街头美式烧烤餐厅
美国，霍普金斯
美式复古风格营造时尚而又亲切的感觉

MILLER'S GUILD 220
SEATTLE, USA
Visual Impression of Icons
磨坊主烤肉餐厅
美国，西雅图
图标的视觉印象

OTHER CREATIVE RESTAURANT 其他创意餐厅

STEBUKLAI 224
VILNIUS, LITHUANIA
An Incredible Baltic Cuisine Restaurant with Wondrous Design and Adventurous Food
奇迹餐厅
立陶宛，维尔纽斯
从奇妙变幻的设计到独特大胆的美食，这是一家不可思议的波罗的海美食餐厅

"BAKERY" 228
VIENNA, AUSTRIA
Independent and Confident Red
"面包房"
奥地利，维也纳
独立而自信的红

KOMBINAT 232
POZNAN, POLAND
LOGO: Not a Purely Mix of Typefaces
联合餐厅
波兰，波兹南
LOGO: 不是字体特效的堆砌

BLACK BEAR 234
BOGOTÁ, COLOMBIA
Vintage Design Reminds You of Your Grandparents' House
黑熊餐厅
哥伦比亚，波哥大
复古设计雕刻出如爷爷家般的温暖时光

OCIO RESTAURANT 238
MEDELLIN, COLOMBIA
Head Chef's Educational Culinary Travels
奥西奥餐厅
哥伦比亚，麦德林
主厨的烹饪求学之旅

THE BARN 240
BERKSHIRE, UK
A Relaxed Atmosphere Created by Healing Illustrations
谷仓餐厅
英国，伯克郡
治愈系插画带出轻松氛围

SUEGRA SABORES CASEROS 242
MEDELLIN, COLOMBIA
Warmth through Bright Colours and Wood
婆婆餐厅
哥伦比亚，麦德林
明亮色彩与木质相契的温暖

GRONBECH AND CHURCHILL 246
COPENHAGEN, DENMARK
Great Colour Combination of Black, White and Gold
格伦比赫与丘吉尔餐厅
丹麦，哥本哈根
黑白金的色彩气质

BARE RESTAURANT 248
BERGEN, NORWAY
Typography: A Movement of Words
贝尔餐厅
挪威，卑尔根
版式设计：文字的乐章

RESTAURANT BERG 250
STUTTGART, GERMANY
Symbolic Imprint of a Wine Glass
伯格餐厅
德国，斯图加特
象征意义的红酒杯印

MEETOWN 254
SHEN ZHEN, CHINA
"Castles, Keys, Old Trees, Animals, Stars, Gardens"
谜堂餐厅
中国，深圳
"城堡，钥匙，老树，动物，星空，花园"

FOODOLOGY 256
SINGAPORE
Visual Features of Pattern Design
美食学餐厅
新加坡
图案设计的视觉特征

PIDGIN 258
SINGAPORE
A Ferryman Who Travels between Now and Past
融合餐厅
新加坡
现代与怀旧的摆渡者

PODI THE FOOD ORCHARD 262
SINGAPORE
Inspired by Spices and Herbs
波蒂美食林
新加坡
灵感汲取于调料与香草

SOCIETY 266
JAKARTA, INDONESIA
Information Visualisation
社会餐厅
印度尼西亚，雅加达
信息的可视化

COOPERS HALL 268
PORTLAND, USA
Traditional Mechanical Effect
桶匠餐厅
美国，波特兰
传统的机械效果

OUTPOST 903 270
SINGAPORE
A Mix of Vintage and Modernity
前哨 903 酒吧
新加坡
复古与现代的混合体

CAFE CHIQUILIN 272
STUTTGART, GERMANY
The Combination of Art Nouveau Fancywork, a Handwriting and a Sans Serif Type to Create a Convivial and Nostalgic Flair
萨尔·库里奥索餐厅
德国，斯图加特
新艺术风格刺绣品与手写无衬线字体打造欢快怀旧氛围

frieze
Volt
14

MEXICAN RESTAURANT

墨西哥餐厅

MONTERREY, MEXICO

ALBERTO SENTIES CATERING

Odd Foil Patch as Regional Symbol

阿尔贝托 · 桑提斯餐饮 / 墨西哥，蒙特雷

具有地域象征的奇怪银箔图形

The design proposal upraises Sentíes' authorship primarily by the use of a seal to communicate his gastronomic mastery. The seal uses a typographical combination of Akzidenz Grotesk and Didot to give it a surprising, elegant and modern touch. The vast play on typography found throughout the brand represents the chef's creative and expansive use of texture in his culinary creations. His style is clinically clean, precise and loyal to its location, the northeastern city of Monterrey in northern Mexico. The business card's odd foil patch is actually a piece of the city's iconic Cerro de la Silla silhouette.

Design agency: Anagrama Client: Alberto Senties Catering

VINO
DE
LA
CASA.

GRA-
CIAS.

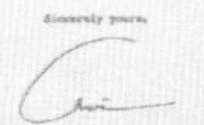

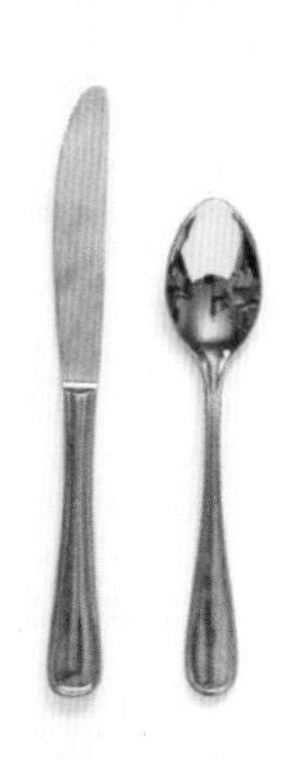

CATERING

设计方案主要利用印章来突出了桑提斯的主厨地位，以此来体现他的品牌形象。印章采用了 Akzidenz Grotesk 和 Didot 两种字体，呈现出一种奇妙、优雅而现代的效果。品牌的字体设计体现了主厨在菜品创作中的创新精神和对食材的广泛应用。他的风格简洁清晰，具有正宗的本地（墨西哥北部城市蒙特雷）特色。名片设计中奇怪的银箔效果其实是该城的地标——西利亚山的剪影。

设计机构：Anagrama 设计公司 委托方：阿尔贝托 · 桑提斯餐饮

SANTIAGO DE QUERÉTARO, MEXICO

EL CARINITO

Dynamic Mexican Handmade Graphics

亲爱的餐厅 / 墨西哥，克雷塔罗

具有活力墨西哥色彩的手绘图形

Collateral design for a Mexican restaurant. The designers wanted to evoke a laid-back marketplace atmosphere in the restaurant. Without shouting out clichés in every element, they focused on the humble elements in regular marketplace kitchens and took inspiration from the way people solves graphic issues by handmade.

Designers: Abraham Lule & Kuro Strada Client: El Cariñito Restaurant

项目是为一家墨西哥餐厅所提供的全套视觉设计。设计师希望在餐厅里营造出一种悠闲的市集氛围，他们没有墨守成规，而是将重点放在普通的市集厨房上，并且从手绘图形中获得了灵感。

设计师：阿布拉汉姆·卢勒、库洛·斯特拉达 委托方：亲爱的餐厅

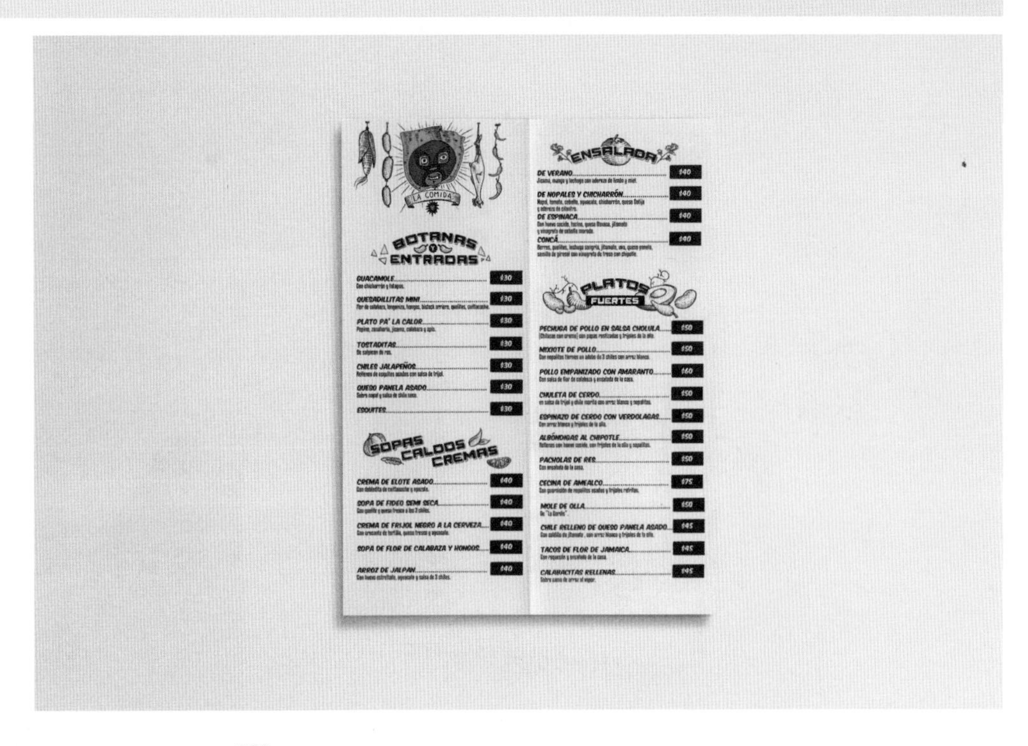

EL
CARINITO
COCINA CON SENTIMIENTO

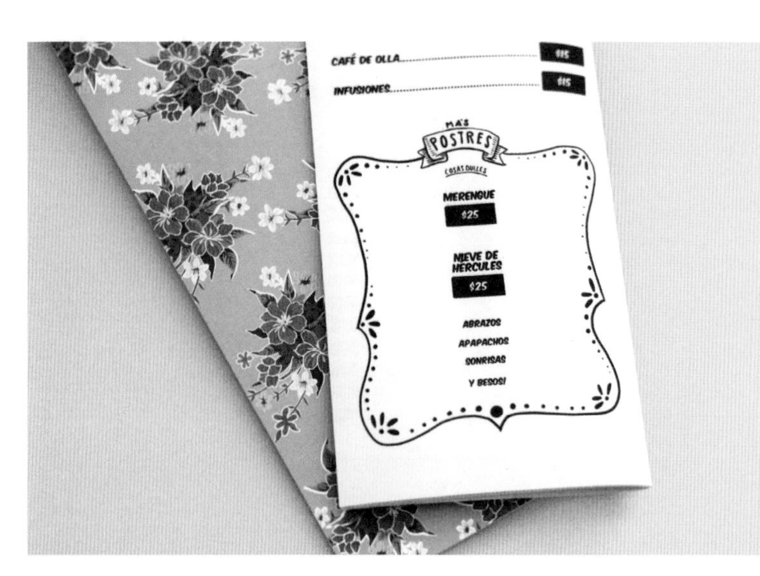
CAFÉ DE OLLA
$15
INFUSIONES
$15
MÁS
POSTRES
COSAS DULCES
MERENGUE
$25
NIEVE DE HERCULES
$25
ABRAZOS
APAPACHOS
SONRISAS
Y BESOS!

EL
CARINITO
COCINA CON SENTIMIENTO

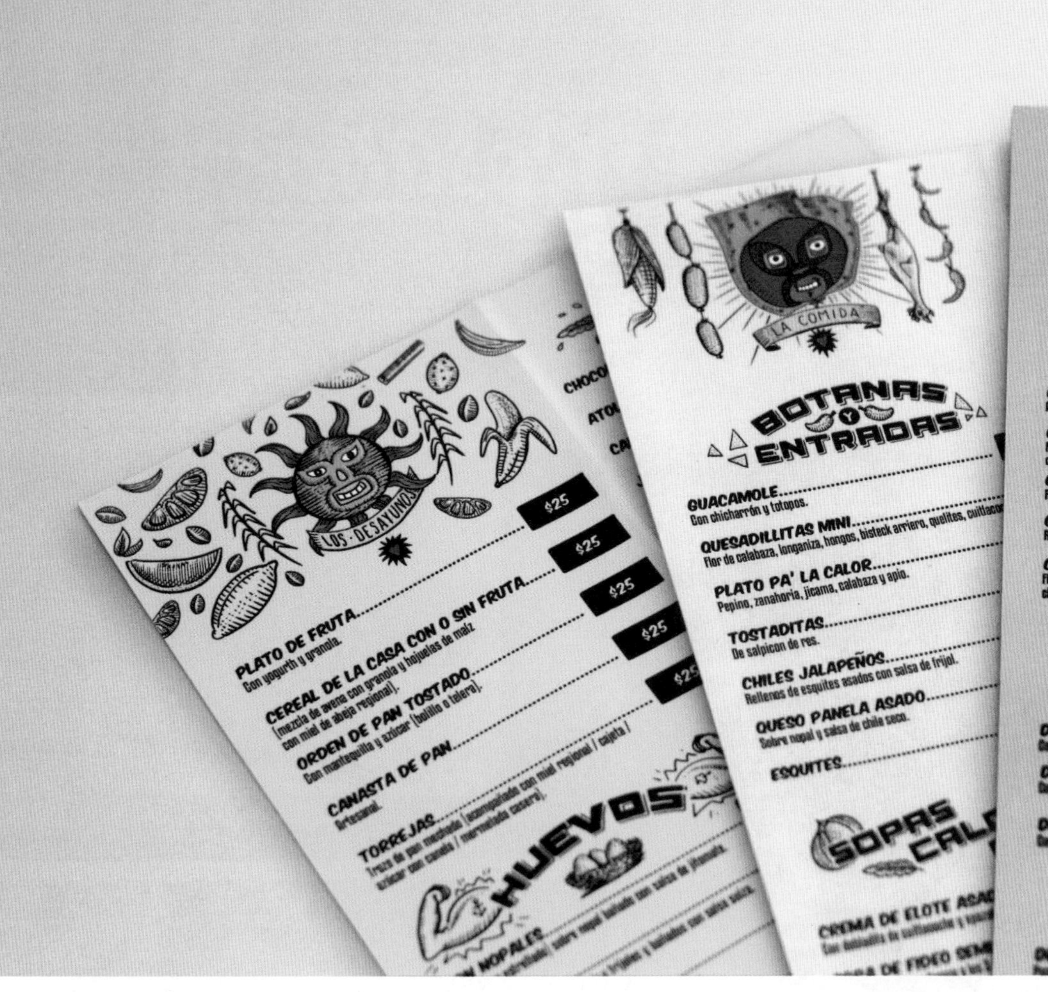
LOS DESAYUNOS
$25
$25
$25
$25
PLATO DE FRUTA
Con yogurth y granola.
CEREAL DE LA CASA CON O SIN FRUTA
(mezcla de avena con granola y hojuelas de maíz con miel de abeja regional).
ORDEN DE PAN TOSTADO
Con mantequilla y azúcar (bolillo o telera).
CANASTA DE PAN
Artesanal.
TORREJAS
HUEVOS
LA COMIDA
BOTANAS Y ENTRADAS
GUACAMOLE
Con chicharrón y totopos.
QUESADILLITAS MINI
PLATO PA' LA CALOR
Pepino, zanahoria, jícama, calabaza y apio.
TOSTADITAS
De salpicón de res.
CHILES JALAPEÑOS
Rellenos de esquites asados con salsa de frijol.
QUESO PANELA ASADO
Sobre nopal y salsa de chile seco.
ESQUITES
SOPAS

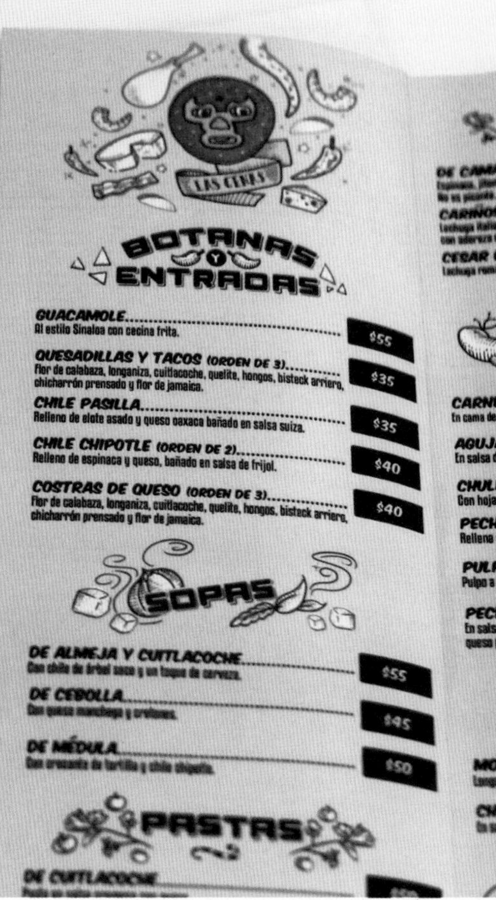
BOTANAS Y ENTRADAS
GUACAMOLE
Al estilo Sinaloa con cecina frita.
$55
QUESADILLAS Y TACOS (ORDEN DE 3)
Flor de calabaza, longaniza, cuitlacoche, quelite, hongos, bisteck arriero, chicharrón prensado y flor de jamaica.
$35
CHILE PASILLA
Relleno de elote asado y queso oaxaca bañado en salsa suiza.
$35
CHILE CHIPOTLE (ORDEN DE 2)
Relleno de espinaca y queso, bañado en salsa de frijol.
$40
COSTRAS DE QUESO (ORDEN DE 3)
Flor de calabaza, longaniza, cuitlacoche, quelite, hongos, bisteck arriero, chicharrón prensado y flor de jamaica.
$40
SOPAS
DE ALMEJA Y CUITLACOCHE
Con chile de árbol seco y un toque de cerveza.
$55
DE CEBOLLA
Con queso manchego y crotones.
$45
DE MÉDULA
Con crocante de tortilla y chile chipotle.
$50
PASTAS
DE CUITLACOCHE

EL
CARIÑITO
COCINA CON SENTIMIENTO

"La Peña del Rosario", is a restaurant located in Puebla, México; a city full of tradition and gastronomical richness. The project is inspired by an enigmatic place. This place is full of life and strange fruits. The designers made a big effort bringing to the customers the best graphic experience.

Designer: José Martín Ramírez Carrasco Photography: José Martín Ramírez Carrasco Client: La Peña del Rosario

罗萨里奥餐厅位于墨西哥普埃布拉——一座具有丰富的传统和美食文化的城市。项目从神秘小屋中获得了灵感，空间中充满了生活气息和奇怪的水果。设计师力求为消费者带来最好的视觉体验。

设计师：何塞·马丁·拉米莱斯·卡拉斯科 摄影：何塞·马丁·拉米莱斯·卡拉斯科 委托方：罗萨里奥餐厅

PUEBLA, MEXICO

LA PEÑA DEL ROSARIO

A Pure and Enigmatic Design

罗萨里奥餐厅 / 墨西哥，普埃布拉

返璞归真的设计质感

MEXICO CITY, MEXICO

PUEBLA 109

Classic Symbols from the Age of Mexican Philately

普埃布拉 109 号餐厅 / 墨西哥，墨西哥城

墨西哥集邮时代的经典符号

Puebla 109 is a new gastronomic hotspot in Roma, the burgeoning DF neighbourhood. Inside the three-floored 20th century townhouse is where art, design and gastronomy converge, in the forms of a restaurant, a bar and a member's club. In the morning the space can be used as a work hub while the space evolves as the day unfolds, offering a nice lunch in the afternoon or a cocktail in the evening.

Design agency: Savvy Studio Photography: Coke Bartrina Client: Puebla 109

The identity for Puebla 109 was developed around several symbols which draw inspiration from the classic age of Mexican philately. Each symbol works independently but at the same time shares an equal hierarchy when used together with the rest of the symbols that make up the graphic system. Unlike a more traditional approach to branding, there is no one symbol that bears the weight of the entire brand's identity.

The applications are constructed upon basic or more industrial materials. They are contrasted with bold colours and classic typefaces that have a strong national character, together with a few other graphic elements which resemble those used by the postal service in the past, therefore alluding to the journey that an object undertakes before reaching its final destination. The interior design was developed by Marcela Lugo and Arturo Dib and includes works of art by Marcos Castro, Lucía Oceguera, Juan Caloca and Luis Alberú.

普埃布拉109号是位于墨西哥城一个新兴社区的一家美食餐厅。在这座建于20世纪的三层联排别墅中，艺术、设计与美食交织在一起，以餐厅、酒吧、会员俱乐部的形式呈现出来。早晨，这里是办公中心；午后，它提供美味的餐饮；晚上，你可以来一杯鸡尾酒。普埃布拉109号的品牌形象围绕着来自于墨西哥集邮时代的几个经典符号展开。各个符号独立运作，同时又共享一个层级，与其他符号共同组成了平面图形系统。与传统的品牌设计不同，这里并没有哪一个符号承载着整个品牌的形象。设计以基本材料甚至工业材料为主。它们与大胆的色彩和具有强烈民族风格的经典字体形成了对比，融合了一些过去邮政服务所使用的图形元素，暗示了邮件在到达目的地之前的一系列经历。室内设计由马塞拉・卢戈和阿图罗・迪布开发，加入了马科斯・卡斯特罗、卢西亚・奥赛格拉、胡安・卡罗卡和路易斯・阿尔贝鲁等艺术家的作品。

设计机构：Savvy工作室 摄影：寇可・巴特里纳
委托方：普埃布拉109号餐厅

STOCKHOLM, SWEDEN

CALEXICO'S

A Mix of Street and Luxury

卡勒西可餐厅 / 瑞典，斯德哥尔摩

街头感爆棚的豪华餐厅

In early 2013 the designers got the assignment from Sweden's most credible and celebrated rock club Debaser to help them brand their new Mexican restaurant. The brief was to create an identity that would feel like a street taqueria gone mad with luxury. Calexico's was made with the ambition to take Mexican food to Stockholm's dining rooms, mixed with Sangria and frozen Margaritas. A fine blend of street tacos, a trendy bar and fine dining. The designers decided to hand paint the type, inspired by Mexican street taqueria signs and they chose a pastel colour palette accompanied by copper to add the luxurious feel. Everything carried the identity – from the menu to the clothes and the plates. The menus were printed as disposable coasters, which helped the hungry who now have their food options available the instant they sit down. A highly skilled window painter helped to add the final touch, actually using real gold on the restaurant windows. When the restaurant opened the identity of Calexico's got a lot of praise, and visitors really liked the mix of street and luxury. It continues to be a well-attended restaurant.

Design agency: SNASK Designers: Jens Nilsson & Magdalena Czarnecki Client: Debaser

2013 年初，设计师受邀为瑞典最著名的摇滚俱乐部 Debaser 新开张的墨西哥餐厅进行全套的品牌营销设计。设计的目标是打造一种豪华版墨西哥街头快餐厅的感觉。卡勒西可餐厅决心将墨西哥美食、桑格里厄汽酒和冷冻玛格丽塔酒带到斯德哥尔摩人的餐桌上，是街头饭馆、时尚酒吧和高档餐厅的完美融合。设计师决定像墨西哥街头饭馆一样使用手绘招牌，以粉彩色系为主色调，辅以铜件装饰来增添奢华感。每个细节都呈现着品牌主题，从菜单到桌布和餐盘。菜单采用一次性纸张印制，方便饥饿的食客能快速点菜。最后，一位技艺高超的橱窗画家为餐厅的窗户配上了真金装饰。在卡勒西可餐厅开业当天，它的品牌设计大获好评，食客们也十分喜爱街头与奢华混合的概念。餐厅一直深受欢迎，热度不减。

设计机构：SNASK 设计公司 设计师：延斯·尼尔森、马格达莱纳·恰尔内茨基 委托方：Debaser 俱乐部

Calexico's
STARTERS
MENY
EN MEXIKANSK
BAR & RESTAURANG
Desserts
MAINS
SOFT TACO
GRINGA
TOSTADA
ALL HOMEMADE!
BAR BROOKLYN
SIDES
RESTAURANGENS ÖPPETTIDER
BRUNCHBUFFÉ

Calexico's
VÄLKOMMEN
TILL EN HET
SOMMAR PÅ
DEBASER
HORNSTULLS
STRAND 4
EN MEXIKANSK
BAR & RESTAURANG
DJS · HELGBRUNCH · UTESERVERING

Calexico's
TAKE
AWAY
EN MEXIKANSK
BAR & RESTAURANG

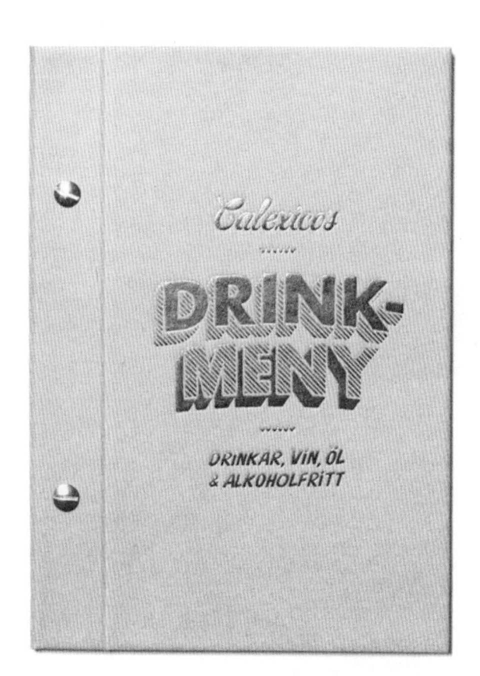
Calexico's
DRINK-
MENY
DRINKAR, VIN, ÖL
& ALKOHOLFRITT

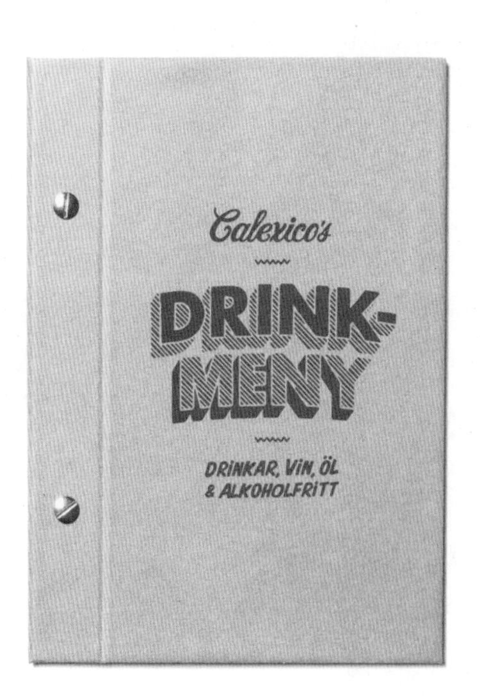
Calexico's
DRINK-
MENY
DRINKAR, VIN, ÖL
& ALKOHOLFRITT

SINGAPORE

MEXOUT

A Mexican Restaurant with Eccentric Thoughts and Design

墨西哥人餐厅 / 新加坡

充满古怪思想与设计的墨西哥餐厅

Mexout is a fresh-mex eatery in Singapore. Imagining Mexout to be a young eccentric Mexican food expert, or "Mex'pert" as the designers coined it, the shop interior is styled to their vision of his living quarters in the basement of his parent's house. As with most eccentric experts, he keeps a wall-of-clues with a Mexican map and pinned locations to track down the freshest ingredients and their suppliers.

Design agency: Bravo Creative director: Edwin Tan Art director: Jasmine Lee Designer: Jasmine Lee Project management: Janice Teo Client: Mexout

Being anti-establishment, Mexout does not adopt a proper brand logo but presents it differently every time. About 20 hand-drawn logos were created to be used in rotation. Similarly for the rest of the collaterals, every element was handwritten or hand-drawn with no use of a computer for the creation of any graphics.

墨西哥人餐厅是新加坡的一家新兴墨西哥餐厅。设计师将餐厅的经营者想象成是一个年轻而古怪的墨西哥美食家，餐厅就设在他父母房子的地下室里。与大多数古怪的专家一样，他在墙壁上挂了一张墨西哥地图，并且用图钉做出标记，以此来探寻最新鲜的食材和它们的供应商。墨西哥人餐厅笃信“反正统主义”，因此并没有采用一个固定的品牌 LOGO，而是每次都以不同的形式呈现。设计师总共设计了 20 款手绘 LOGO 供餐厅轮流使用。餐厅的周边产品设计同样全部采用手写或手绘装饰，没有使用一点电脑图形。

设计机构：Bravo 设计公司 创意总监：艾德文·谭 艺术总监：杰斯敏·李 设计师：杰斯敏·李 项目管理：珍妮丝·泰奥 委托方：墨西哥人餐厅

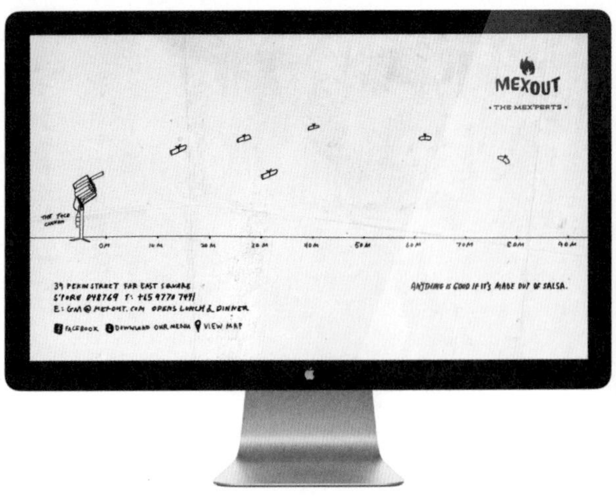

MEXOUT

MEXOUT
PICK YOUR STYLE!
BURRITO
TACO
RICE BOWL
SALAD BOWL
NACHOS
MEXOUT
TACO ANATOMY
Mexout
MEXOUT

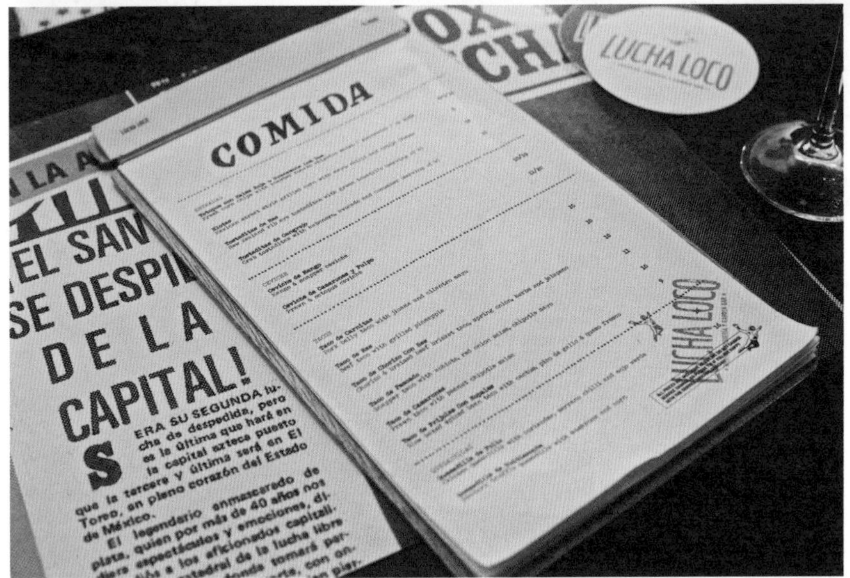

Meet Lucha Loco, Singapore's first Mexican taqueria which meets two critical needs in your life – gourmet Mexican street food and crazy Mexican wrestling. Bravo set out to design the brand by not designing much. The objective was to make everything look raw and natural as though the bar has existed forever. The namecards mimic vintage wrestling trading cards.

Design agency: Bravo Creative director: Edwin Tan Art director: Amanda Ho Designer: Amanda Ho Client: Lucha Loco

摔跤餐厅是新加坡第一家既能享用墨西哥街头美食又能领略疯狂的墨西哥摔跤的墨西哥餐厅。Braovo 设计公司决定以极简主义为设计主题，目标是让一切都呈现出原生态的感觉，让人以为餐厅已经存在很久了。餐厅的名片模仿了传统的摔跤商业名片。

设计机构：Bravo 设计公司 创意总监：埃德温·谭 艺术总监：艾曼达·霍 设计师：艾曼达·霍 委托方：疯狂摔跤餐厅

SINGAPORE

LUCHA LOCO

Minimalism Provides an Original Feeling

疯狂摔跤餐厅 / 新加坡

极简主义赋予的“原生态”感觉

Casa Virginia is Chef Mónica Patiño's latest culinary project, where she looks to glorify and reenact a homely experience. The designers encompass Casa Virginia as the easiness and familiarity of eating at home, with the highest quality in its cuisine. The same philosophy and ideals are reflected upon the graphic applications which, though simple in nature, express an upmost attention to the smallest details – through special finishes such as gold foiling – imitating the chef's meticulous process in her cooking. The identity was developed as a contemporary reinterpretation of the traditional graphic language that was popular in Mexico during the 1920s.

Design agency: Savvy Studio Client: Casa Virginia

弗吉尼亚私房菜馆是主厨莫妮卡·帕提诺最新的美食餐厅，她试图给人以高雅的家庭就餐体验。设计师对这家餐厅的定位是"提供至上美味和轻松环境的私房菜馆"。图形设计反映了同样的设计定位，本质简单的图形通过金箔等特殊装饰将重点放在微小的细节上，模仿了主厨一丝不苟的烹饪流程。餐厅的品牌形象以现代的方式重新诠释了墨西哥 20 世纪 20 年代的流行图形语言。

设计机构：Savvy 工作室 委托方：弗吉尼亚私房菜馆

MEXICO CITY, MEXICO

CASA VIRGINIA

Detail Is a Sort of Language

弗吉尼亚私房菜馆 / 墨西哥，墨西哥城

细节也是一种语言

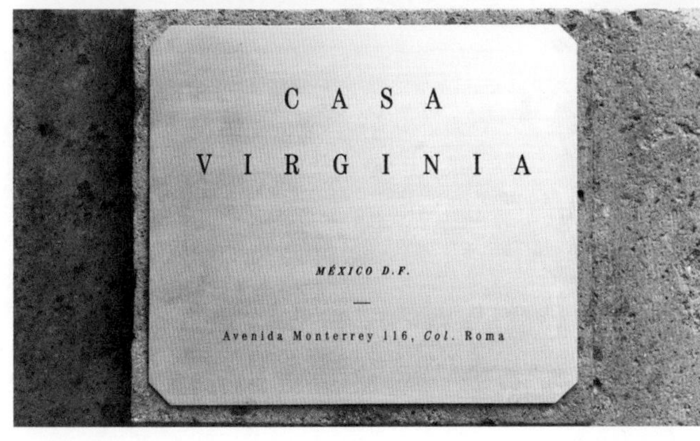

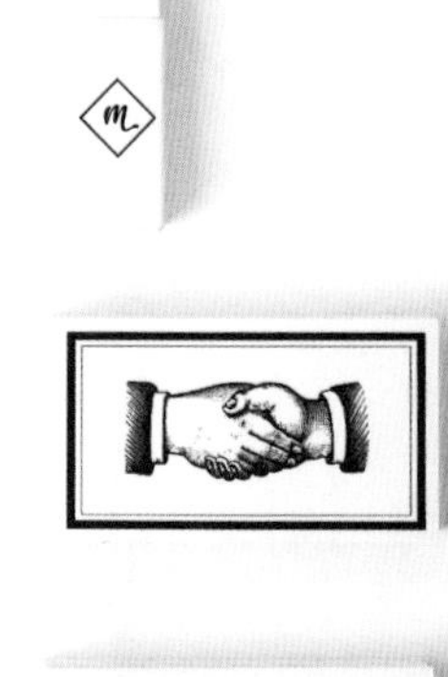

Visual identity project for a Spanish-Mexican restaurant in Mexico City. The design reflects the use of rare spices, creative dishes, handmade type and unique illustrations creating harmony between ingredients, taste and design. Montagü is born from the idea of Chef Roger Weber. Shaping in a menu, the experience that his travels around the world gifted him. Collecting flavors and ingredients from the corners of Asia, Africa, Europe and Mexico, with a unique character, Roger has achieved that his creations become a journey of flavors for Montagü's guests.

Design agency: NHOMADA Designer: Diego Leyva Photography: Diego Leyva Client: Montagü Gastro Winebar / Rogelio Weber

项目是为墨西哥城一家西班牙 / 墨西哥风味餐厅所进行的视觉形象设计。设计反映了餐厅对罕见调料的应用、创意菜品、手工制作等特色，通过独特的插画在食材、品味和设计之间形成了完美的平衡。蒙塔古餐厅的设计概念来自于主厨罗杰·韦伯，环游世界的经历让他受益匪浅。他从亚洲、非洲、欧洲和墨西哥的各个角落收集各种调料和食材，让蒙塔古的食客在餐厅中享受美味之旅。

设计机构：**NHOMADA 设计公司** 设计师：**迪亚戈·雷瓦** 摄影：**迪亚戈·雷瓦** 委托方：**蒙塔古餐厅 / 罗杰·韦伯**

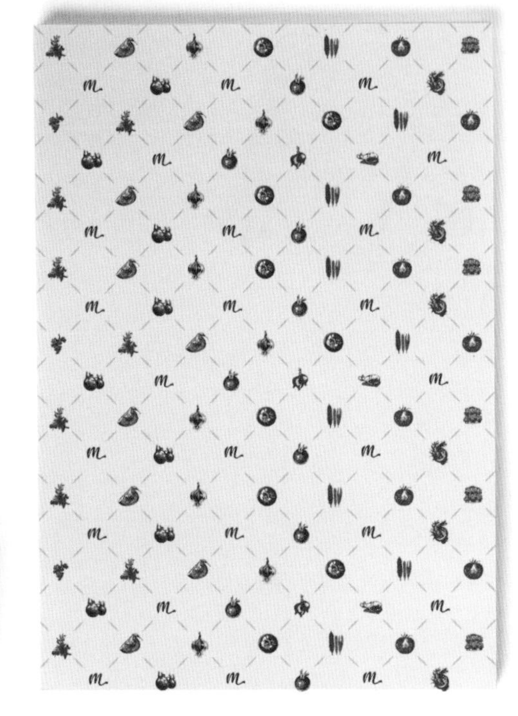

MEXICO CITY, MEXICO

MONTAGU

Design Inspiration from Travel Experience around the World!

蒙塔古餐厅 / 墨西哥，墨西哥城

设计灵感来源于主厨环游世界的经历！

SPANISH RESTAURANT
西班牙餐厅

OURENSE, SPAIN

TAMARINDO

Adding Fresh and Soft Ice-cream Colours to a Restaurant with Rainy Galician Qualities

塔马林多餐厅 / 西班牙，奥伦塞

为仿若加利西亚阴雨连绵般气质的餐厅添加一抹清新柔和的冰淇淋色

Located in Ourense, Spain, this place is anything but typical, is a refreshing alternative for local walkers used to the traditional bar or restaurant. The architect Ruben D. Gil and his wife Gretta R. Valdés decided to spice up the rainy Galician city with an unusual spot to enjoy international cuisine and drinks in an atmosphere of light wood ceilings, adobe walls, dim lighting and steel furniture. The agency worked on the design of its visual identity, stationery materials, packaging for branded products, to go packaging, coasters, menus and tote bags. A custom made bottle of water on each table welcomes its guests in October 2014.

Design agency: La Tortillería Designers: Zita Arcq, Rodrigo Véjar Client: Gretta Valdés and Ruben Gil

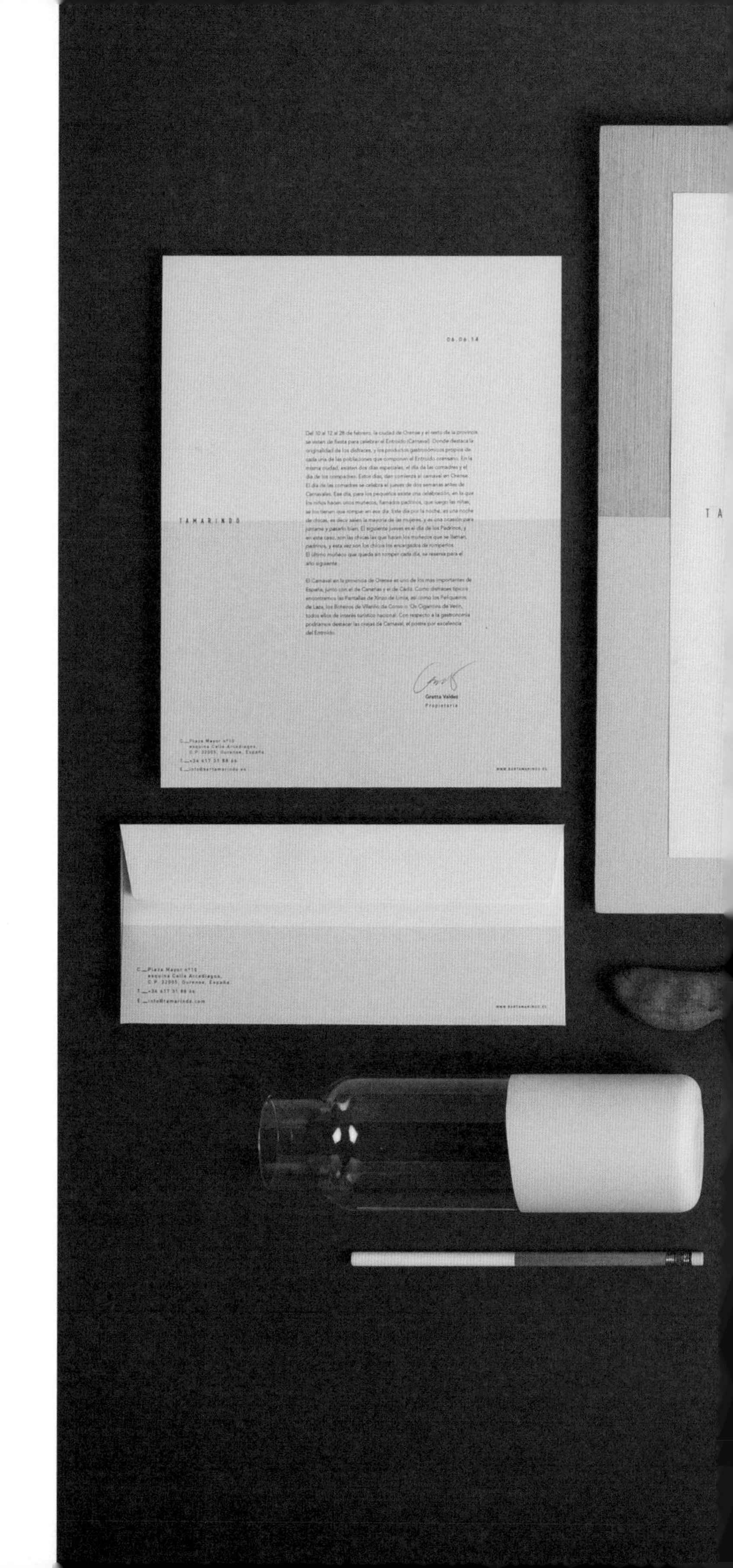

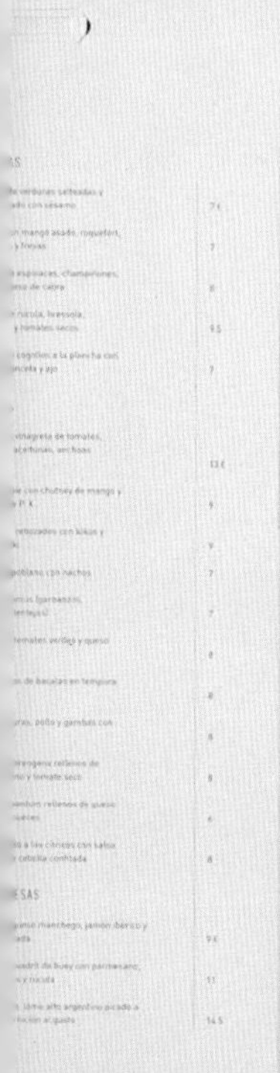

TAMARINDO
A

TAMARINDO
TAMARINDO

TAMARINDO
PATIO BAR

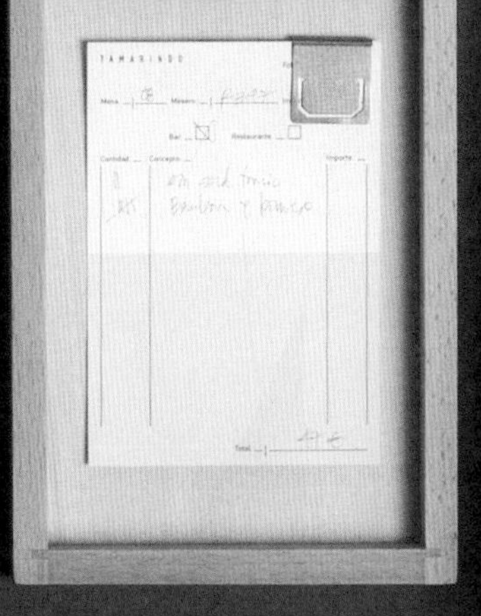
TAMARINDO

TAMARINDO
CASA COCINA
- Ensalada con mango asado, roquefort, frutos rojos y fresas
- Ensalada de espinacas, champiñones, nueces y queso de cabra
- Ensalada de rúcula, bresaola, parmesano y tomates secos
- Ensalada de cogollos a la plancha con refrito de panceta y ajo
PLATILLOS
- Burrata con vinagreta de tomates, almendras, aceitunas, anchoas y albahaca
- Mi-cuit de foie con chutney de mango y reducción de P.X.
- Langostinos rebozados con kikos y salsa Teriyaki
- Guacamole poblano con nachos
- Surtido de humus (garbanzos, berenjena y lentejas)
- Tempura de tomates verdes y queso de cabra
- Fish and chips de bacalao en tempura con alli-oli
- Wok de verduras, pollo y gambas con arroz jazmín
- Rollitos de berenjena rellenos de queso pecorino y tomate seco
- Raviolis de wantum rellenos de queso de Arzúm y nueces
- Sticks de pollo a los cítricos con salsa de mostaza y cebolla confitada
HAMBURGUESAS
- Casera: con queso manchego, jamón ibérico y cebolla confitada
- Tradicional: cuadril de buey con parmesano, tomates secos y rúcula
- Angus Iversen: lomo alto argentino picado a cuchillo, guarnición al gusto
7€
7
8
9.5
7
13€
9
9
7
7
8
8
8
8
6
8
9€
11
14.5

TAMARINDO
TAMARINDO
TAMARINDO
TAMARINDO

TAMARINDO
TAMARINDO

这家位于西班牙奥伦塞的餐厅不同于传统的酒吧或餐厅，给人以耳目一新的感觉。建筑师鲁本·D·吉尔和他的妻子格里塔·R·瓦尔德斯决定为这座阴雨连绵的加利西亚城市带来一个可以享用各国美食和饮品的非凡场所，这里有浅木色天花板、土砖墙壁、昏暗的灯光和钢铁家具。La Tortillería 设计公司对餐厅的视觉形象、文具材料、产品包装、打包袋、杯垫、菜单和购物袋进行了全套设计。每张餐桌都配有特别定制的水瓶，餐厅于 2014 年 10 月正式开张。

设计机构：La Tortillería 设计公司 设计师：思蒂·阿尔克、罗德里格·维加尔 委托方：格里塔·瓦尔德斯、鲁本·吉尔

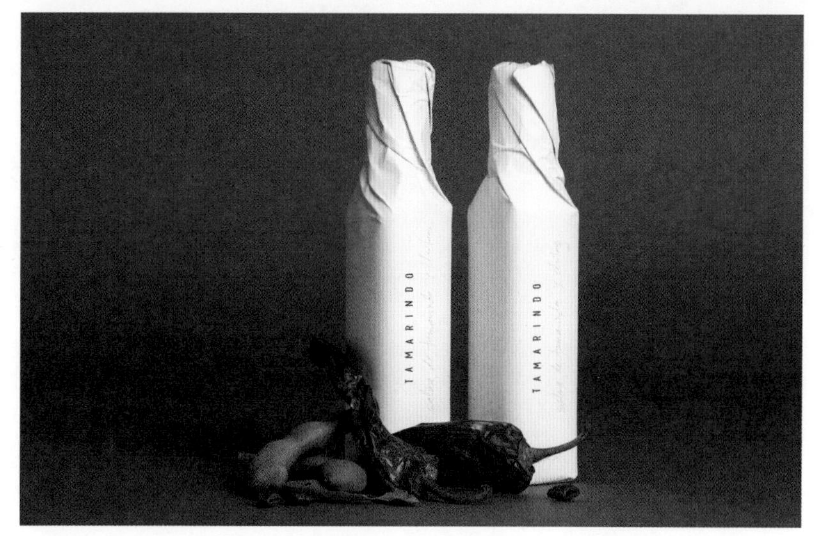

OSLO, NORWAY

VINO VERITAS

Design Is as Natural as "Ripening of Grape"

维利塔斯葡萄酒餐厅 / 挪威，奥斯陆

设计应该是像"葡萄成熟了"一样自然的事

Designing the brand identity and interior design for a Spanish ecologic gastrobar in Oslo (Norway), the studio Masquespacio was inspired by the craftsmanship and ecology that determines the brand values of Vino Veritas. That's why a vintage image was chosen to recuperate the traditional way of elaborating things. In first case, it can be noticed that the main logo contains a vine leaf and some grapes to highlight the core business of Vino Veritas. On the other hand the packagings created for the different products like olive oil, wine and almonds just use leafs. A classic Arabic pattern is another aspect to highlight through a reinvention from which a rhombus is extracted like an icon for the different communication tools. Further on major importance is given to ecology using recyclable paper and reusable packagings like the aged one for almonds that can easily be reused for storage in the kitchen by Vino Veritas' customers. Last but not least 'handmade' is stood out through the paper, cord and scissors through packaging. "It was important for us to highlight characteristics like the ecology and craftsmanship of Andalusia that were marked significantly by the business model of Vino Veritas. Although we wanted to represent it through an image easily acceptable for a wider public, looking for the Spanish culture and gastronomy with an ecologic and handmade character."

Design agency: Masquespacio Creative director: Ana Milena Hernández Palacios Photography: David Rodríguez and Carlos Huecas Client: Vino Veritas

VINO VERITAS
PAELLA
VINO VERITAS
ECO-GASTROBAR
Glass + tapa
Flaske + 2 tapas
Sangría + tapas
OLIVO

10 BASIC MEDITERRANEAN
DIET RECOMMENDATIONS
Use olive oil as your main source of added fat.
Eat plenty of plant-based foods. For dessert, eat fresh fruit on a daily basis.
Bread and other cereal foods should be a part of your everyday diet.
Fresh, locally produced and in-season foods that have undergone minimal processing are best.
Consume dairy products on a daily basis.
Red meat should be consumed in moderation.
Eat plenty of fish, but eggs in moderation.
Water is the beverage par excellence in the Mediterranean. Wine should be taken in moderations and with meals.
Be physically active every day.
Be physically active every day.
THIS IS OUR STORY
We turned to organic farming and its natural and healthy lifestyle after some health problems in our family, caused by intensive crop cultivation practices.
We found our haven in the Alpujarra region of the province of Almeria (South East Spain), and El Cortijo El Cura became one of Spain's pioneering organic farms.
Situated in Laujar de Andarax, a village steeped in history and tradition, our family business aims to develop sustainably within the framework of this rich local legacy.
The views from our farm are breathtaking, the magical sunsets over the Sierra Nevada mountain range painting a background to unique and unforgettable Mediterranean flavours.
Our farm, our philosophy and our pleasure. We hand you over to your experience.
The Sánchez Vizcaino family.

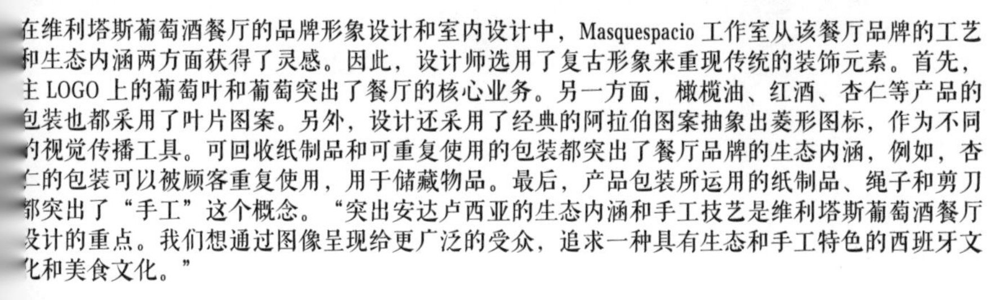

在维利塔斯葡萄酒餐厅的品牌形象设计和室内设计中，Masquespacio 工作室从该餐厅品牌的工艺和生态内涵两方面获得了灵感。因此，设计师选用了复古形象来重现传统的装饰元素。首先，主 LOGO 上的葡萄叶和葡萄突出了餐厅的核心业务。另一方面，橄榄油、红酒、杏仁等产品的包装也都采用了叶片图案。另外，设计还采用了经典的阿拉伯图案抽象出菱形图标，作为不同的视觉传播工具。可回收纸制品和可重复使用的包装都突出了餐厅品牌的生态内涵，例如，杏仁的包装可以被顾客重复使用，用于储藏物品。最后，产品包装所运用的纸制品、绳子和剪刀都突出了“手工”这个概念。“突出安达卢西亚的生态内涵和手工技艺是维利塔斯葡萄酒餐厅设计的重点。我们想通过图像呈现给更广泛的受众，追求一种具有生态和手工特色的西班牙文化和美食文化。”

设计机构: Masquespacio 设计公司 创意总监: 安娜·米莱娜·赫尔南德·帕拉西奥斯 摄影: 大卫·罗德里格斯、卡洛斯·韦卡斯 委托方: 维利塔斯葡萄酒餐厅

An inventive Spanish culinary offering called for a brand story upheld by intricate illustrations to personify the identity of Sal Curioso, an inventor obsessed with the science of cooking. His culinary experiments are illustrated observations which permeate the restaurant, as each piece of designed communication purposefully reflects Sal's unique character.

Design agency: Substance Executive creative director: Maxime Dautresme Designers: Siuming Leung, Hinz Pak, Belinda Alfonso, Vicky Lum, Julia Pak, Claudia Yuen, Anna Ceccenamo Client: Woolly Pig Concepts

这家别出心裁的西班牙餐厅希望用插画来呈现它的品牌故事，将萨尔·库里奥索塑造成了一个沉浸于研究烹饪技术的发明家。他的烹饪实验以插画的形式呈现在餐厅的各个角落，每件插画作品都反映了萨尔的独特个性。

设计机构：Substance 设计公司 创意总监：马克西姆·多特里斯米 设计师：秀明·梁、辛斯·派克、博林达·阿方索、维奇·卢、茱莉亚·派克、克劳迪娅·袁、安娜·塞斯纳莫 委托方：Woolly Pig Concepts 公司

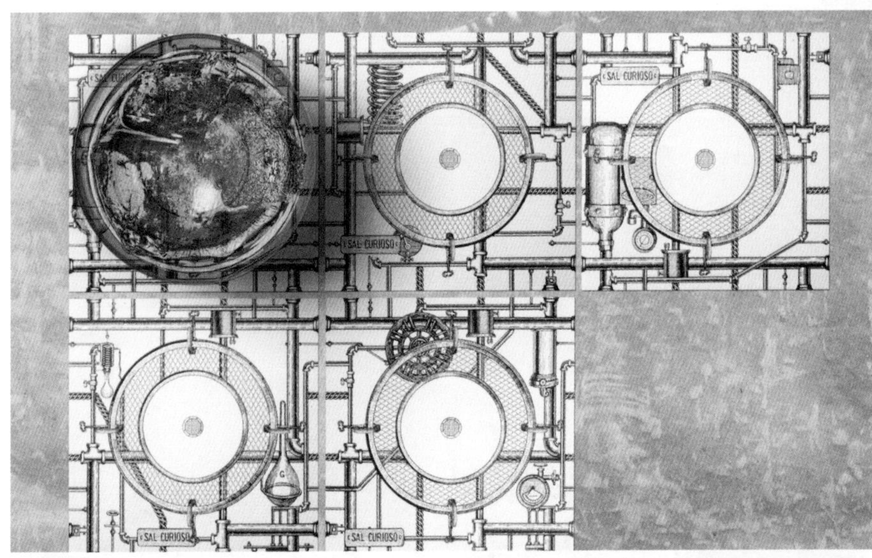

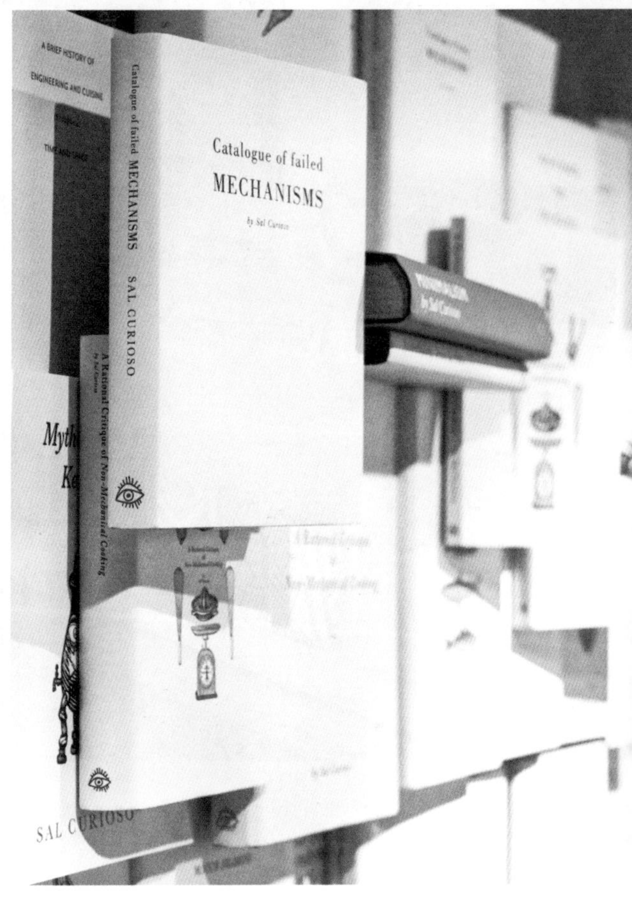

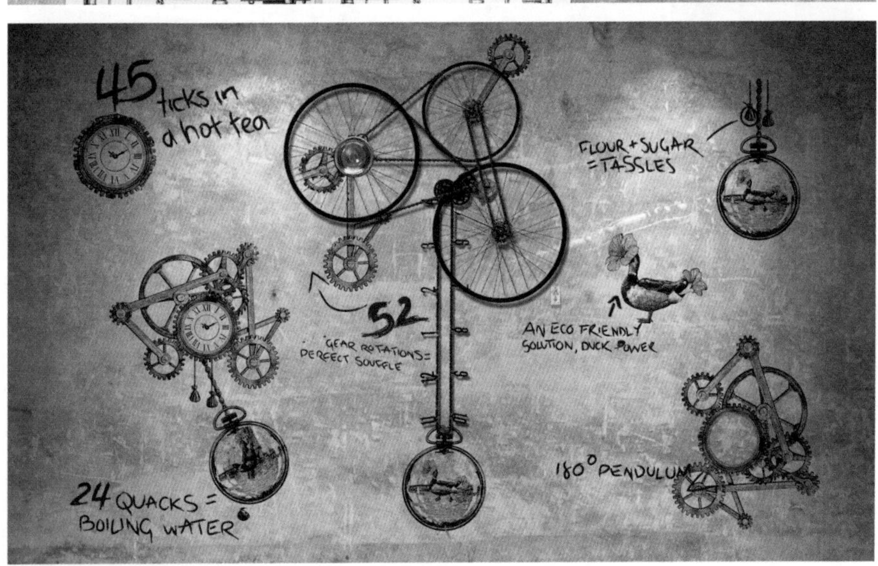

HONG KONG, CHINA

SAL CURIOSO

To Show Unique Culinary Experiments in Illustrations

萨尔·库里奥索餐厅 / 中国，香港

以插画形式呈现独特烹饪实验

La cocote
Siedfleisch mit Meerrettichsauce,
Bouillonkartoffeln und Gemüse
Le plat de jour
Kabeljau mit Jakobsmuscheln,
Limonensauce & Spinatrisotto
Les légumes
Orecchiette mit Gemüse & Bärlauchpesto

FRENCH RESTAURANT

法式餐厅

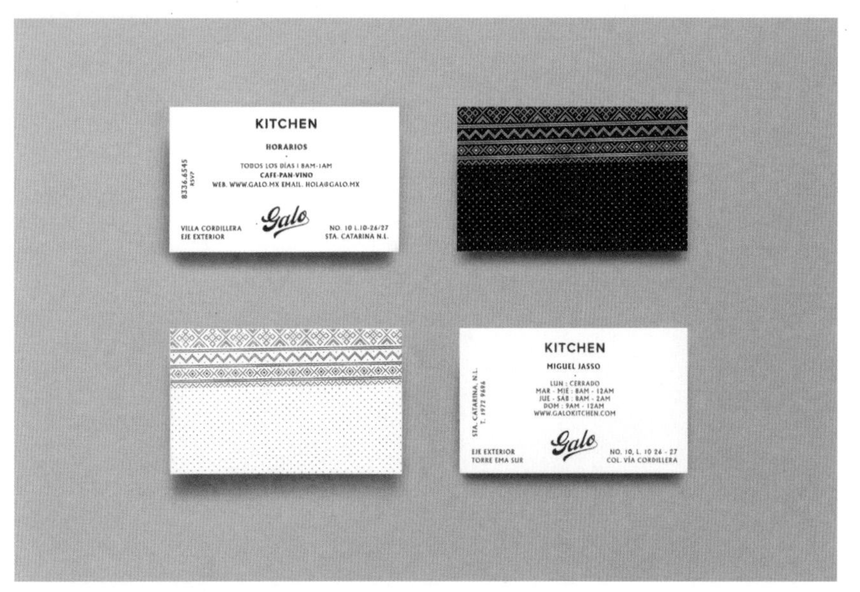

MONTERREY, MEXICO

GALO

Homage from Zeppelin Icon

加洛小厨 / 墨西哥，蒙特雷

齐柏林式飞艇图标的敬意

Galo Kitchen is a restaurant specialising in French-American inspired comfort cuisine. Its prime focus is breakfast, but it also offers lunch and dinner menus and a cozy atmosphere all day long. Additionally, Galo Kitchen has its own in-house bakery that provides delicious, freshly made, hand-crafted bread and pastries. The naming is meant to articulate the French touch present in Galo's lovingly made food. The black and white skewed pattern dresses up the brand as friendly, snug and casual, a feeling supported by the logotype's organic cursive script. The zeppelin icon pays homage to Galo's bakery. Inspired on the airship's general shape, a Zeppelin is a sandwich made with a roll of French bread split widthwise into two pieces and filled with a variety of meats, cheese, vegetables, seasonings and sauces.

Design agency: Anagrama Designer: Lucía Elizond

Galo
KITCHEN
HORARIOS

Galo
KITCHEN
HORARIOS
TODOS LOS DIAS I 8AM-1AM
CAFE-PAN-VINO
WEB. WWW.GALO.MX EMAIL. HOLA@GALO.MX
8336.4545
RSVP
Galo
VILLA CORDILLERA
EJE EXTERIOR
NO. 10 L.10-26/27
STA. CATARINA N.L.

Galo
KITCHEN

AM
Galo
PM
KITCHEN
Galo
PALE ALE
Galo
KITCHEN
Galo
OLIVAS
Galo
MERMELADA DE FRAMBUESA
MENU
BRIE
Galo
CAMEMBERT
QUESO DE LA CASA
Galo
Galo

Galo
MERMELADA DE FRAMBUESA
Galo
CAFÉ
BOLD
HOUSE BLEND

MENÚ
FRUTAS Y PANES
HUEVOS Y SALADOS
Galo

加洛小厨是一家专注法美融合美食的餐厅。餐厅以早餐著称，但是也提供午餐和晚餐，整体氛围舒适融洽。此外，加洛小厨还有自己的烘焙坊，为人们提供美味新鲜的手作面包和糕点。餐厅的命名具有浓厚的法国风情。黑白两色的倾斜图案给予了品牌友好、温馨、轻松的感觉，而 LOGO 的草写字体则加重了这种感觉。齐柏林式飞艇图标的设计旨在向加洛烘焙坊致敬。齐柏林式飞艇三明治的造型与飞艇类似，由两块法式面包卷和各种肉类、芝士、蔬菜、调味品和酱汁制成。

设计机构：Anagrama 设计公司 设计师：卢西亚·艾利桑德

NEWBURY, UK

THE WOODSPEEN

Michelin Star Chef John Campbell's New Venue: Simple Elegance

伍德斯宾餐厅 / 英国，纽伯里

米其林星级主厨的新餐厅：简洁的优雅

The idea for Michelin star chef John Campbell's new venture was fine dining for everyone. & SMITH wanted the logotype to be uncomplicated and simple, the identity to avoid the usual clichés you'd expect from fine dining, and the website to lead with suppliers and the people to make it happen (rather than keeping them in the background).

Design agency: & SMITH Designer: Sam Kang Photography: & SMITH

米其林星级主厨约翰·坎贝尔的新餐厅，力求将高级餐厅的体验推广到所有人中间。& SMITH 希望餐厅的 LOGO 简洁明快，脱离典型的高级餐厅形象，设计将供应商和餐厅工作人员从幕后带到了台前，给人以耳目一新之感。

设计机构：& SMITH 设计公司 设计师：山姆·康 摄影：& SMITH 设计公司

SINGAPORE

BALZAC

Humanist Consciousness of a French Restaurant

巴尔扎克餐厅 / 新加坡

一家法式餐厅的人文情怀

Identity design for an authentic French restaurant in Singapore. Concept of the brand is based loosely on French novelist and playwright Honoré de Balzac. A quill and inkwell make up the icon of the logo. The designers handpicked a few of Balzac's amusing quotes, those with references to food and beverages, and placed them around the interior of the restaurant in appropriate typographical treatment. The designers also created a couple of posters inspired by Balzac's novels to play up the concept.

Design agency: Bravo Creative director: Edwin Tan Designers: Amanda Ho, Pharaon Siraj Client: Balzac Brasserie

本项目是为新加坡一家正宗的法式餐厅所提供的形象设计。该餐厅的品牌理念以法国著名作家巴尔扎克为基础，餐厅的LOGO由羽毛笔和墨水瓶构成。设计师精心挑选了一些巴尔扎克与餐饮相关的经典语录，经过合适的文字排版处理，将它们放置在餐厅的各个角落。设计师还以巴尔扎克的小说为灵感，制作了一些切合主题的海报。

设计机构：Bravo设计公司 创意总监：埃德温·谭 设计师：阿曼达·霍、法拉翁·西拉杰 委托方：连锁反应项目公司

BALZAC

BALZAC

BALZAC
BRASSERIE
PETIT DÉJEUNER
PLATS
DU JOUR
LE DINER

BALZAC
BRASSERIE
BALZAC

Échine
Jambon
Épaule
Grillades
Poitrine fraiche
» salée
» fumée
Palette
Lard gras frais
» salé
Pointe
Foie
Rognon
Oreille
Chair à saucisses
Petit salé
Tripes
Saindoux
Choucroute crue
» cuite
Pieds crus
» panés
» campagne 18,50
» viande
Andouillette
Boudin blanc
» noir
Cervelas
Fromage de tête
Galantine
» de volaille
Hure
Jambon blanc
» fumé
Jambonneau
Mortadelle
Rillettes
Rôti porc cuit
Salami
Saucisses fraiches
BALZAC

ZURICH, SWIZERLAND

QUALINOS

Inspiration from French Pot "Le Creuset"

加里诺斯餐厅 / 瑞士，苏黎世

法式“酷彩”锅带来的灵感

Quaglinos is a French brasserie designed by Dyer-Smith Frey. From the business cards through the matches and aprons the Graphic design draws its lines from the typically French pot "Le Creuset" used and illustrated as the main logo. The menus also draw inspiration from the flair and design of a typical French brasserie by illustrating the different meal types.

Design agency: Dyer-Smith Frey Designers: James Dyer-Smith, Gian Frey Photography: Dyer-Smith Frey Client: Kramer Gastronomie

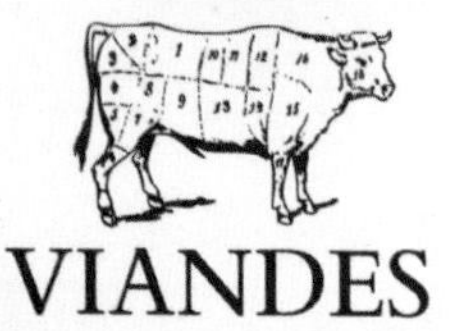

VIANDES

SUPRÊME DE MAIS-POULARDE À L'ESTRAGON *Maispoularde mit Estragon*	32.00
FILET D'AGNEAU AUX HERBES POTAGÈRES *Lammrücken im Kräutermantel*	42.00
ENTRECÔTE „CAFÉ DE PARIS"	46.00
CÔTE DE VEAU *Kalbskotelett*	49.00

POISSONS

Moules Marinières *Miesmuscheln*	32.-
Saumon *Lachs*	34.-
Lotte et sa vierge d'herbes aux tomates et olives *Seeteufel mit Tomaten-Olivensalsa*	44.-
Bouillabaisse	48.-

GARNITURES

Pommes Allumettes *Streichholzkartoffeln*	7.00
Pommes Nouvelles *Neue Kartoffeln*	7.00
Nouilles fines *Feine Nudeln*	7.00
Croustillant de risotto *Knusprige Risottorolle*	7.00
Légumes de la saison tièdes *Lauwarmes Saisongemüse*	7.00
Salade de la saison *Saisonsalat*	7.00

POUR FINIR

Glaces et sorbets «Maison» *Hausgemachte Eissorten*	5.-
Crème Brûlée	12.-
Profiteroles	12.-
Tarte Tatin avec glace vanille *Gestürzter Apfelkuchen mit Vanilleeis*	15.-
Moelleux chocolat *Schokoladenkuchen mit flüssigem Kern*	15.-
Sélection de fromages *Käseauswahl*	18.-

Déclaration des viandes · Fleischdeklaration

Volailles: France, Suisse	Bœuf: Suisse, Australie	Veau: Suisse	Porc: Suisse, Espagne	Agneau: Australie
Geflügel: Frankreich, Schweiz	Rind: Schweiz, Australien	Kalb: Schweiz	Schwein: Schweiz, Spanien	Lamm: Australien

加里诺斯餐厅是一家由 Dyer-Smith Frey 设计公司所设计的法式啤酒屋。从名片、火柴到围裙，餐厅的图形设计从典型的法式“酷彩”锅中获得了灵感，并以其作为餐厅的主 LOGO。菜单还从经典法式啤酒屋的设计中获得了灵感，以插画的形式展示了各种不同的菜品。

设计机构：Dyer-Smith Frey 设计公司 设计师：詹姆斯·戴亚-史密斯、吉安·弗雷 摄影：Dyer-Smith Frey 设计公司 委托方：Kramer 餐饮公司

ZURICH, SWIZERLAND

LOUIS

Stylish Simplicity

路易斯啤酒屋 / 瑞士，苏黎世

有格调的简约

Brasserie Louis is a modern brasserie for which Dyer-Smith Frey created the interior concept and the corporate identity. The strong corporate identity runs consistently throughout each element of the project, bringing them all together. The "Louis" logo appears on plates, espresso cups, napkins and mineral water carafes as a reassuring sign that every piece bearing it stands for the best quality.

Design agency: Dyer-Smith Frey Designer: James Dyer-Smith, Gian Frey Photography: Dyer-Smith Frey
Client: Kramer Gastronomie

路易斯啤酒屋是一家现代啤酒屋，Dyer-Smith Frey 为其提供给了室内概念和企业形象设计。强烈的企业形象贯穿了项目的各个角落，帮助它们融为一体。餐厅“LOUIS”字样的 LOGO 出现在餐盘、咖啡杯、餐巾和矿泉水瓶上，每一个标有 LOGO 的物件都代表着最高的品质，给人以安心之感。

设计机构：Dyer-Smith Frey 设计公司 设计师：詹姆斯·戴亚-史密斯、吉安·弗雷 摄影：Dyer-Smith Frey 设计公司 委托方：Kramer 餐饮公司

FERMO, ITALY

FRU FRU

A Combination of Traditional French Kitchen Values and Homemade Cooking

法鲁餐厅 / 意大利，费尔莫

传统法式烹饪与家常料理在设计中的意境融彻

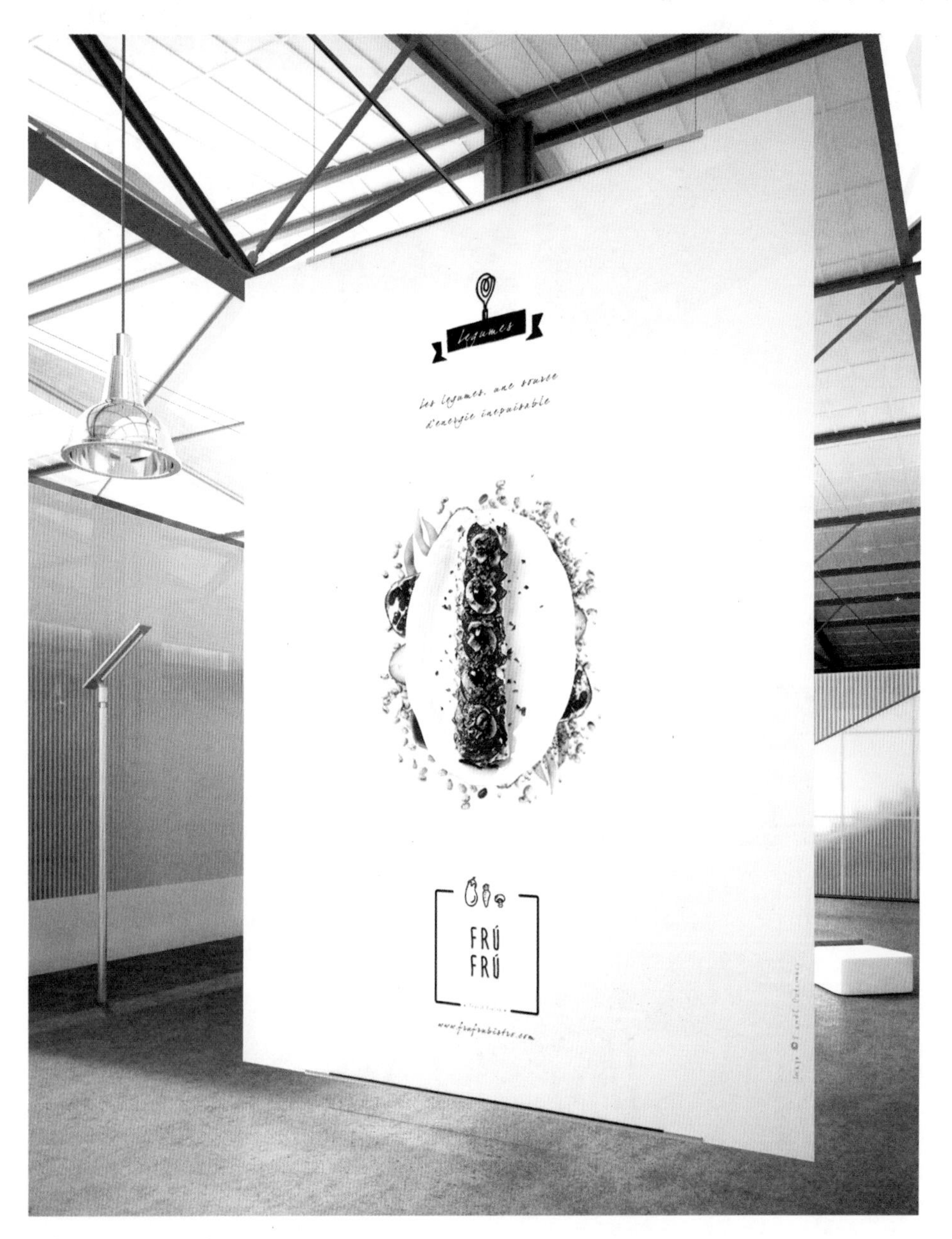

Frú Frú is an Italian restaurant where you can eat fresh meats, vegetables and fruits in a French bistro's style. Frú Frú's menu is inspired by French gastronomy, using high quality Italian ingredients and integrating elements from contemporary kitchens. The designer's job as a brand developer was to create an identity where the restaurant could glorify the traditional French kitchen values and emphasised the quality of an homemade cooking, using fresh local products.

Designer: Johanna Roussel Photography: B. and E. Dudzinscy, Boris Ryzhkov, Jacques Palut, Joshua Resnick Client: Frú Frú

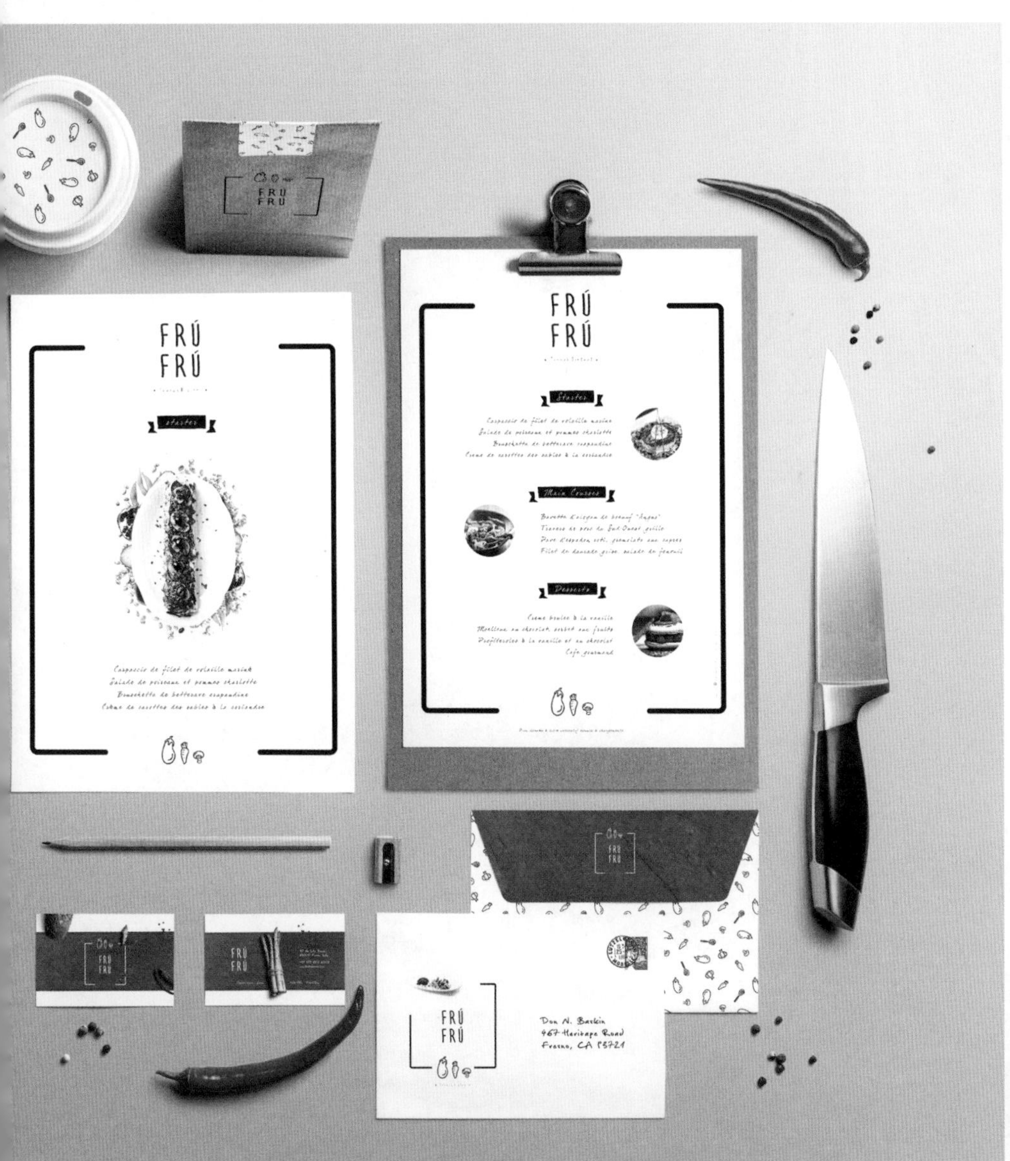

在这家位于意大利的餐厅，你可以体验正宗的法式啤酒屋，尽情享用各种新鲜的肉类、蔬菜和水果。法鲁餐厅的菜单从法国大餐中获得了灵感，选用高品质意大利食材，并且融入了现代厨房的新鲜元素。设计师作为品牌开发者的任务是为餐厅打造一个合理的品牌形象，既要尊重传统法式烹饪的价值，又要突出家常料理和新鲜食材的品质。

设计师：乔安娜·鲁塞尔 摄影：B. 杜新西、E. 杜新西、波利斯·雷日科夫、雅克·帕拉特、乔舒亚·雷斯尼克 委托方：法鲁餐厅

THE LADIES' PURSE
HAPPY VALLEY RACECOURSE
WEDNESDAY, FEBRUARY 25TH, 1863
HOTSPUR v. YEDDO
A Purse with 21 Sovereigns
PRESENTED BY MISS POLLARD
The Unique Prize
PONY ENTRANCE FEE $5

BRITISH RESTAURANT

英伦餐厅

LONDON, UK

THE COLLECTION

Ultimate Fashion in Whiting

收藏餐厅 / 英国，伦敦

清水抹白的极致时尚

The Collection is restaurant, cultural event and retail space. Mind Design designed the identity, signage system and all printed material. The idea for the identity relates to multiple prints, limited editions and artist signatures. The execution is relatively simple: Everything is based on an A5 format with punched holes. The designers used screen printing which allowed them to change colours on the printing bed and makes each print unique. Larger signs are made up by several A5 boards and the thickness is achieved by hanging several signs in front of each other. For the logo the designers asked the client to write the name in their own handwriting connecting two dots equivalent to the punched holes.

Design agency: Mind Design Client: The Collection

the collection

the collection

264 Brompton Road
London SW3 2AS
United Kingdom

+44 (0)20 7225 1212
+44 (0)20 7225 1050
info@the-collection.co.uk
www.the-collection.co.uk

the collection

264 Brompton Road
London SW3 2AS
United Kingdom

+44 (0)20 7225 1212
+44 (0)20 7225 1050
info@the-collection.co.uk
www.the-collection.co.uk

e collection

264 Brompton Road
ondon SW3 2AS
ited Kingdom

+44 (0)20 7225 1212
+44 (0)20 7225 1050
info@the-collection.co.uk
www.the-collection.co.uk

Registered Office
London SW3 2AS,
Registered

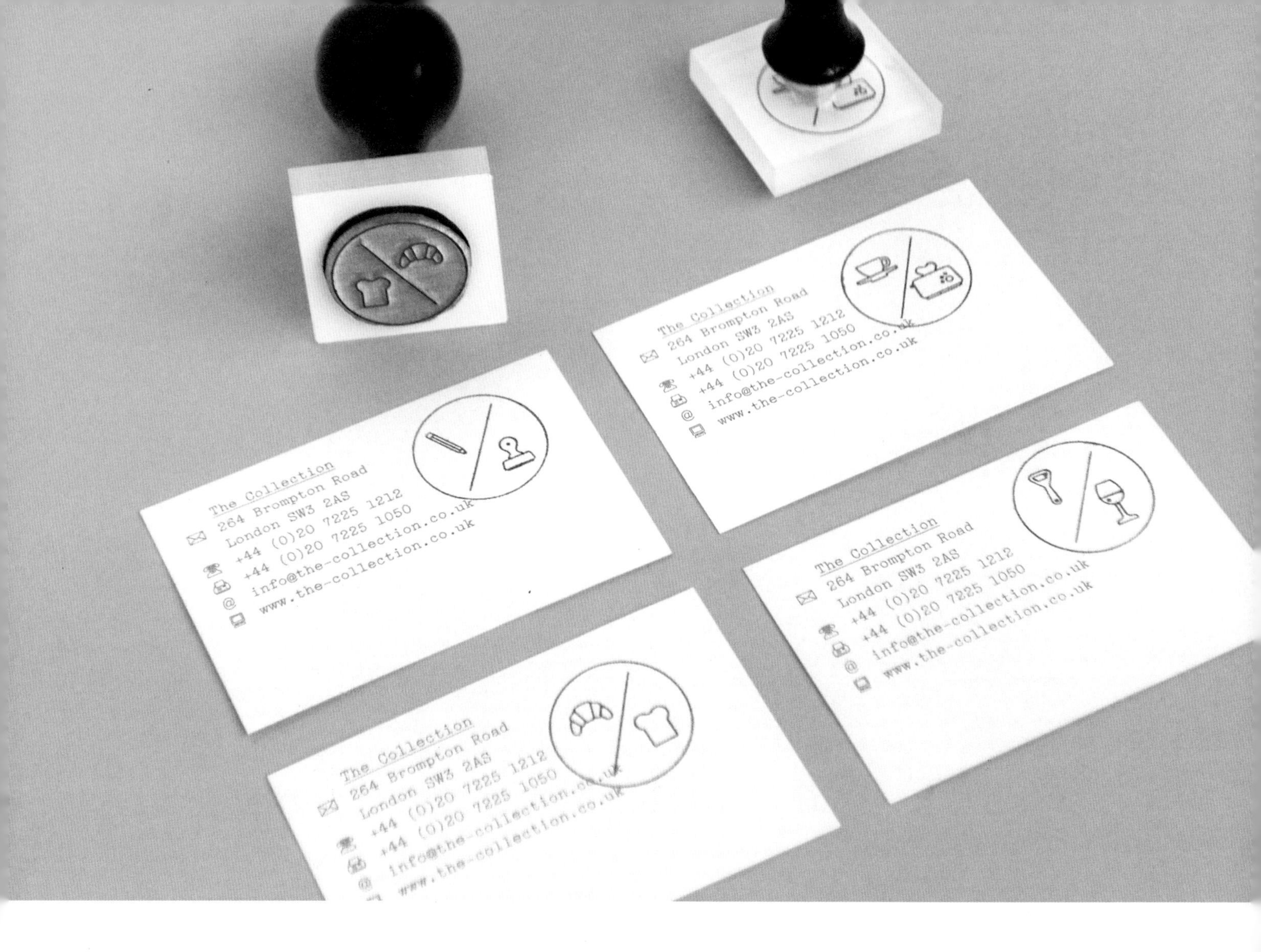
The Collection
264 Brompton Road
London SW3 2AS
+44 (0)20 7225 1212
+44 (0)20 7225 1050
info@the-collection.co.uk
www.the-collection.co.uk

DRINKS MENU
BRUNCH MENU
DAY MENU
DINNER MENU

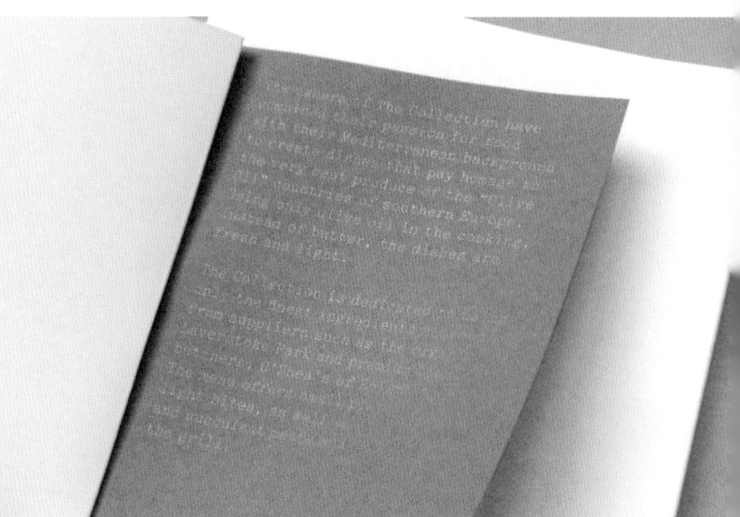

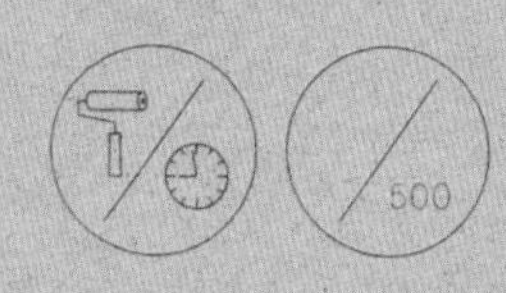

收藏餐厅不仅仅是一家餐厅，还是一个文化场所和零售空间。Mind 设计公司为其设计了品牌形象、导视系统以及所有印刷材料。品牌形象的设计参考了多种印刷品、限量版图书和艺术家的签名。最终执行的设计相对简单：一切都以一张带有打孔的 A5 纸为基础。设计师采用了丝印技术，从而可以在印刷版上改变色彩，让每个印记都独一无二。大型标志由若干张 A5 纸板构成，若干个标志的叠加形成了一定的厚度。在 LOGO 设计中，设计师让客户手写下餐厅的名字，其中的两个点就相当于打孔。

设计机构：Mind 设计公司 委托方：收藏餐厅

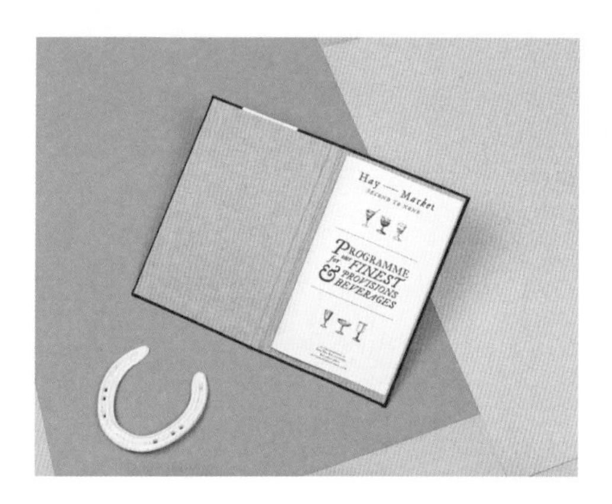

HONG KONG, CHINA

HAY MARKET

Vintage British Style Created by Geometric Shapes and Classic Letterforms

干草市场餐厅 / 中国，香港

几何图形与古典字体打造复古英伦风

Hay Market is a restaurant set in the sprawling grounds of the Hong Kong Jockey Club. With Hong Kong Jockey Club's pedigree as a British Colonial entity, the basis of the brand personality and language is British Eccentricity. Inspired by vibrant jockey silks which are drenched in centuries of tradition and superstition, the restaurant's visual language is an eclectic mix of bold geometric shapes juxtaposed against vintage British typography and Victorian illustrations from old advertisements. The brand's logo is a playful update on classic letterforms and also functions as a blank canvas, allowing for quirky permutations when combined with different illustrations.

Design agency: Foreign Policy Design Group Creative director: Yah-Leng Yu Designers: Liquan Liew, Vanessa Lim, Yah-Leng Yu

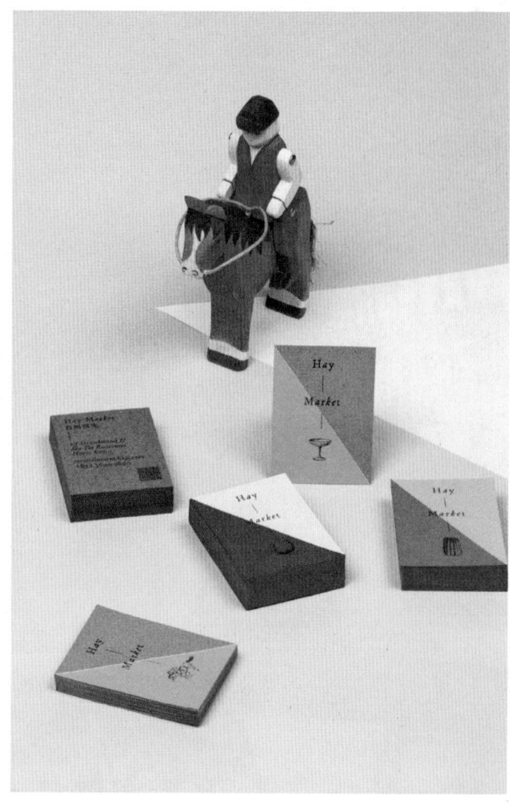

干草市场是位于香港赛马会地块的一家餐厅。由于香港赛马会具有英国殖民气息，餐厅品牌形象的也以英伦风格为基础。餐厅的视觉设计从沉淀了上百年的赛马骑师服装中获得了灵感，混合了各种大胆的几何形状、复古英伦字体和维多利亚时代老广告风格的插画。品牌 LOGO 巧妙地升级了古典字体，同时也形成了留白，让不同的插画形成奇妙的排列组合。

设计机构：Foreign Policy 设计集团 创意总监：亚玲·于 设计师：礼泉·刘、瓦内萨·李、亚玲·于

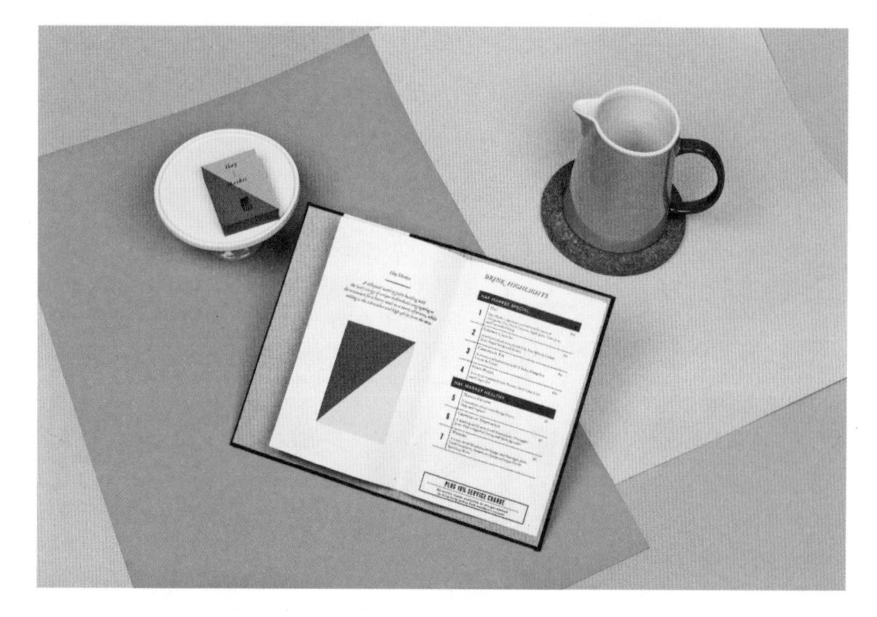

THE LADIES' PURSE
HAPPY VALLEY RACECOURSE
WEDNESDAY, FEBRUARY 25TH, 1863
HOTSPUR v. YEDDO
A Purse with 21 Sovereigns
PRESENTED BY MISS POLLARD
The Unique Prize

THE ACE
SUITE

WASH
ROOMS
PRIVATE
UP
STAIRS
ROOMS
TO LIFTS

NO 2
SECONDI
Merluzzo
fänkål & forellrom
240:-
Brasato al Barolo
185:-
Porchetta di Malatino
24-timmars bakad spädgris porchetta,
215:-

ITALIAN RESTAURANT

意大利餐厅

SINGAPORE

SOPRA

Imaginations of Post-war Italy

索普拉餐厅 / 新加坡

战后意大利的影像

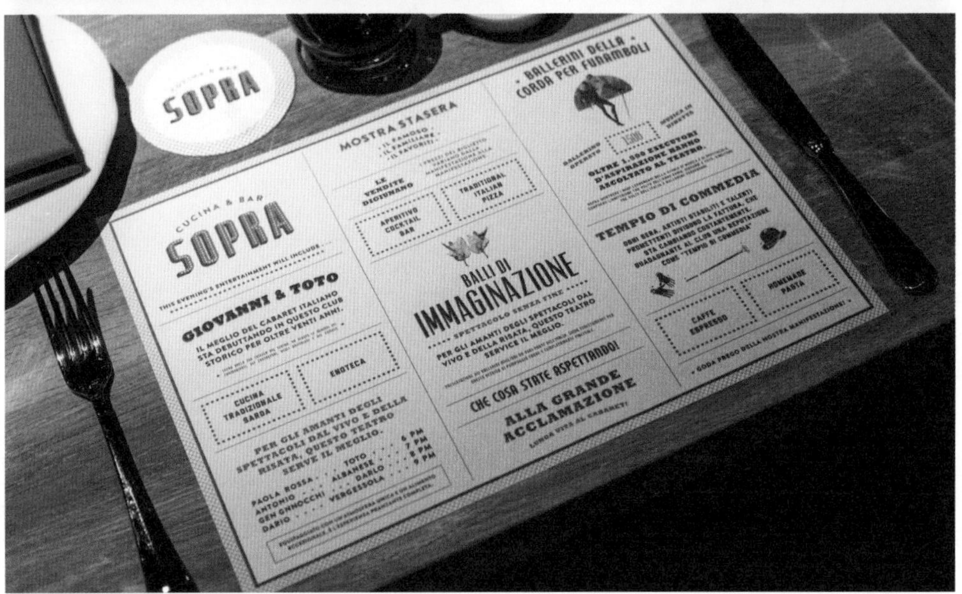

Bravo were commissioned to rebrand Sopra, an Italian restaurant, for their Singapore outlet. The new Sopra Cucina & Bar is an ode to the glamorous days of post-war Italy, when Hollywood and the films of Federico Fellini and Sophia Loren first captured the imaginations of an enamored public.

Design agency: Bravo Creative director: Edwin Tan Designer: Jasmine Lee Client: Sopra

Bravo 设计公司受委托为这家位于新加坡的意大利餐厅重塑品牌形象。全新的索普拉餐厅是一首对战后意大利辉煌岁月的赞歌，费德里科·费里尼和索菲亚·罗兰在影片中第一次将这些影像呈现在公众面前，餐厅的设计正是以此为主题。

设计机构：Bravo 设计公司 创意总监：埃德温·谭 设计师：贾思敏·李 委托方：索普拉餐厅

The Sicilian is an old-style Italian restaurant located in New South Wales, Australia. As the brand name suggests, the design borrows heavily from the gentlemen's style of 1940's gangster films. Bravo imaginesthis to be a place where mafia bosses patronise. The typography of the logo is inspired by engravings on firearms.

Design agency: Bravo Creative director: Edwin Tan Art director: Amanda Ho Designer: Amanda Ho Project manager: Janice Teo

西西里餐厅是一家位于澳大利亚新南威尔士的老式意大利餐厅。正如餐厅的名字所示，它的设计借鉴了 20 世纪 40 年代黑帮电影中的绅士风格。Bravo 设计公司将餐厅想象成一个由黑手党老大保护的场所。LOGO 的字体设计从枪支的刻字上获得了灵感。

设计机构：Bravo 设计公司 创意总监：埃德温·谭 艺术总监：阿曼达·霍 设计师：阿曼达·霍 项目经理：詹尼斯·陶

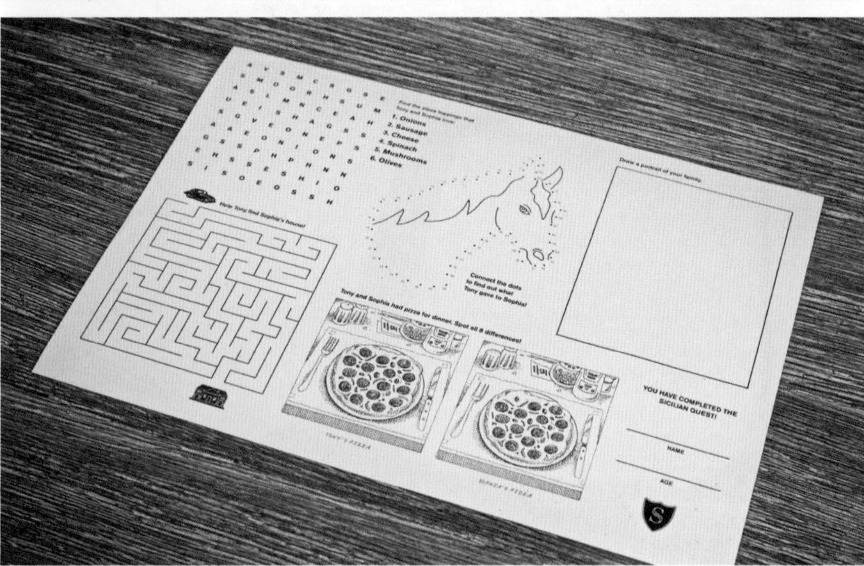

New South Wales, Australia

THE SICILIAN

The Gentlemen's Style of 1940's Gangster Films

西西里餐厅 / 澳大利亚，新南威尔士

20 世纪 40 年代黑帮电影的绅士风格

UN POCO
E-MAIL info@unpoco.se
www.unpoco.se
ADRESS
11431 Stockholm
Emilio Collocolo 0704-565999
TEL 08-6110269

UN POCO

UN POCO

UN POCO
UN POCO

UN POCO

STOCKHOLM, SWEDEN

UN POCO

New York's Rawness VS Swedish Simpleness

一点点餐厅 / 瑞典，斯德哥尔摩

纽约的狂野碰撞瑞典的简约

This classic and luxurious Italian restaurant is inspired by New York's rawness and the Swedish simpleness which pervades in both interior design, graphic identity and taste, offering a clean yet strong typographic expression. Its own house wine is also sold in liquor stores in Sweden. Newly opened Un Poco Bar is located next door.

Design agency: Acid and Marble Designers: Mia Askerstam Nee and Antonio Vergara Alvarez Photography: Acid and Marble Client: Un Poco Restaurant

这家经典奢华的意大利餐厅混合了纽约的狂野和瑞典的简约，将其融入了室内设计、品牌形象和菜品口味之中。餐厅的文字设计简洁而强烈。餐厅自制的红酒目前在瑞典的各大酒类专卖店有售。新开张的一点点酒吧就位于餐厅隔壁。

设计机构：Acid and Marble 设计公司 设计师：米亚·阿斯科尔斯塔姆·尼、安东尼奥·维加拉·阿尔瓦雷斯 摄影：Acid and Marble 设计公司 委托方：一点点餐厅

MONTERREY, MEXICO

IANNILLI

"Romantic Nostalgia" Concept Created by Delicate Craftsmanship

伊安尼里餐厅 / 墨西哥，蒙特雷

精致工艺打造“浪漫怀旧”主题风格

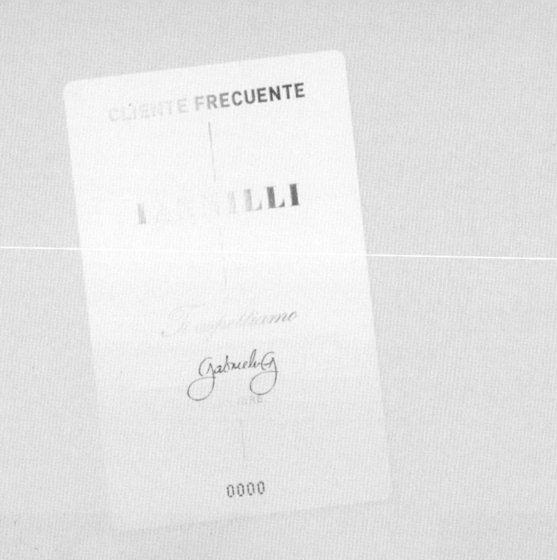

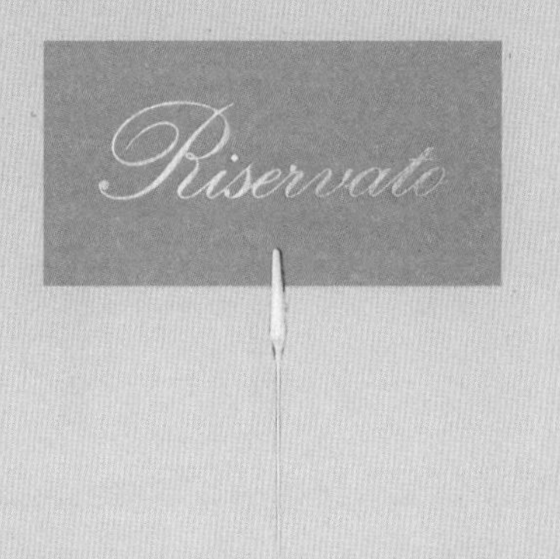

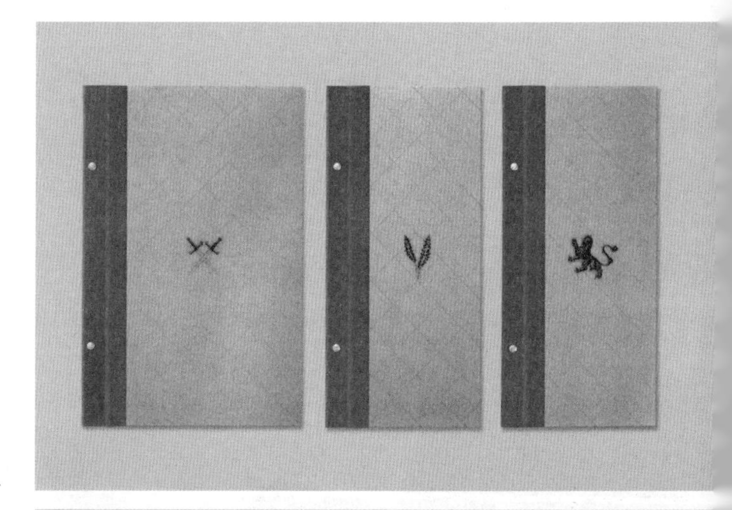

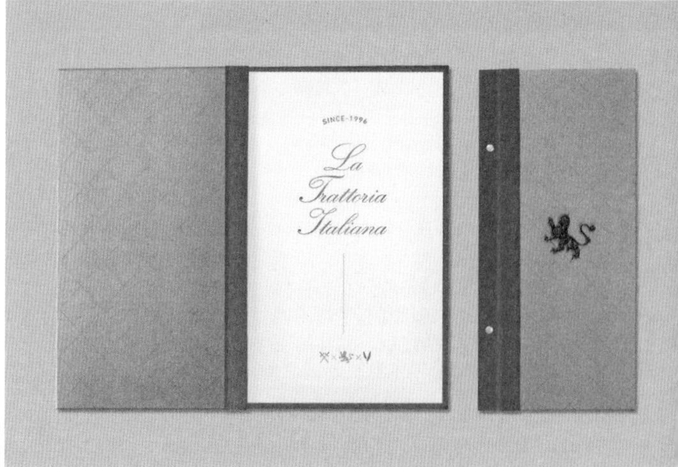

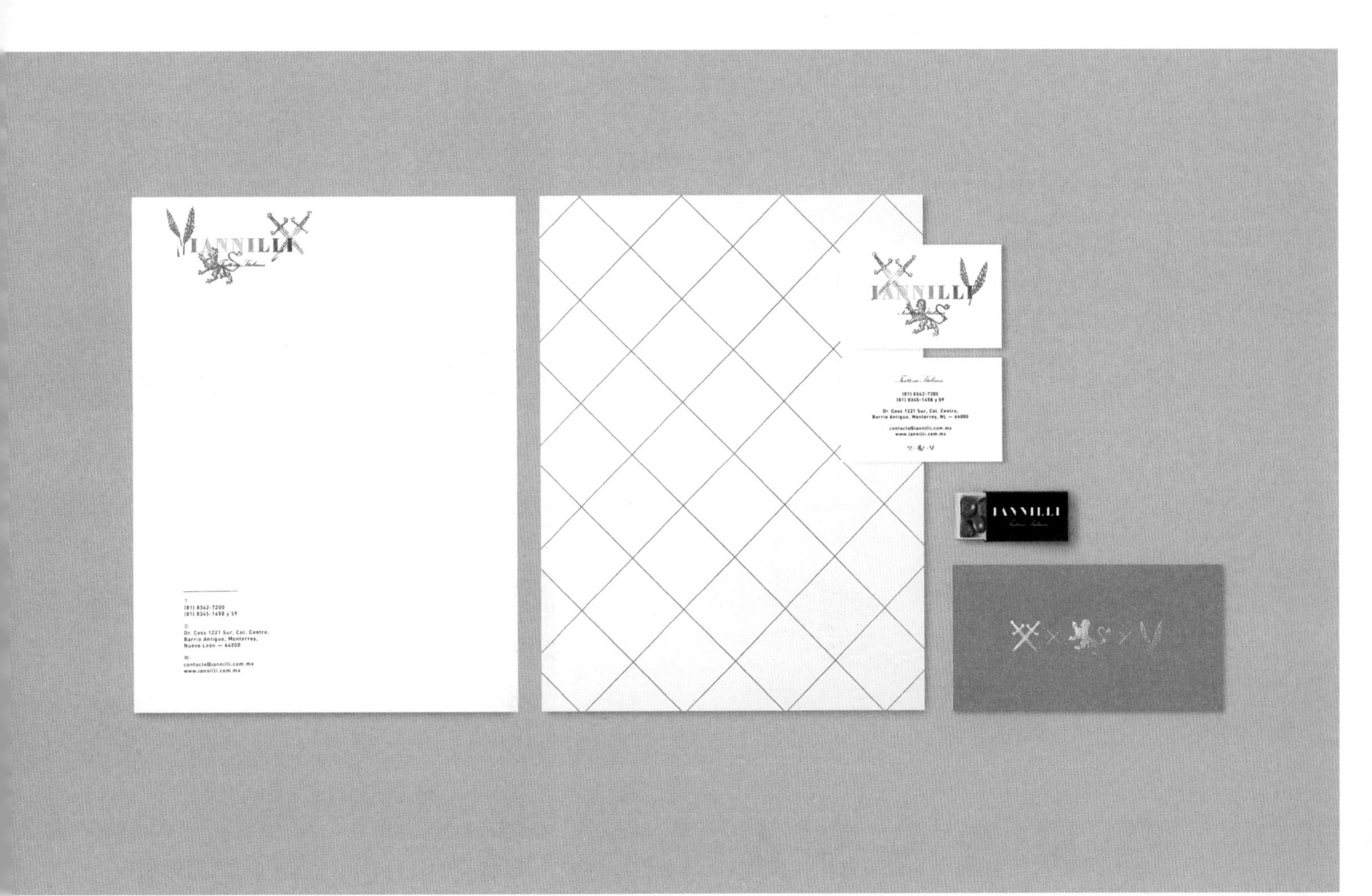

Vith this project Savvy Studio proposed the creation of a jraphic identity that gives honor to the Iannilli family, their estaurant and their loyal clientele that has accompanied them rom the beginning, and, at the same time, to develop an dentity attractive to a younger segment, fueling the growing nterest of young people of providing an important space in heir lives to the dining experience. Following the concept of he "Romantic Nostalgia", Iannilli is defined as an intimate lace to evoke memories and enjoy the romance of a good linner. Tradition, experience and Iannilli's past, at the service f their present guests.

)esign agency: Savvy Studio Photography: Alejandro Cartagena

avvy 工作室希望通过全新的形象设计致敬伊安尼里家族、他们所经营的餐厅以及伴随他们一路走来的忠实顾客；同时，设计师还希望新开发的品牌形象能吸引更多的年轻人，为他们带来特别的就餐体验。以“浪漫还旧”为主题，餐厅被设计成一个私密的场所，给人以浪漫的美食体验。传统、体验和伊安尼里的过去都体现在餐厅的服务中。

设计机构：Savvy 工作室 摄影：亚力山卓·卡塔赫纳

Grassa is a modern Italian restaurant in Portland, Oregon specialising in handmade pasta within a unique environment that balances industrial details with meticulously crafted dishes.

Design agency: Public-Library Designers: Ramón Coronado and Marshall Rake Photography: Dina Avila, David L. Reamer Client: Grassa

格萨拉餐厅是一家位于波特兰的现代意大利餐厅，专营手工意大利面。餐厅独特的环境在工业化的细节设计和精致的菜品之间形成了平衡。

设计机构：Public-Library 设计公司 设计师：拉蒙·科罗纳多、马歇尔·雷克 摄影：蒂娜·阿维拉、大卫·L·雷默 委托方：格拉萨餐厅

PORTLAND, USA

GRASSA

Perceptual Feeling of Commercial Illustration

格拉萨餐厅 / 美国，波特兰

商业插画的直观情感

PIZZA
SUSHI
CAFÉ
ESTE
OESTE
BELÉM - LISBOA

PIZZA

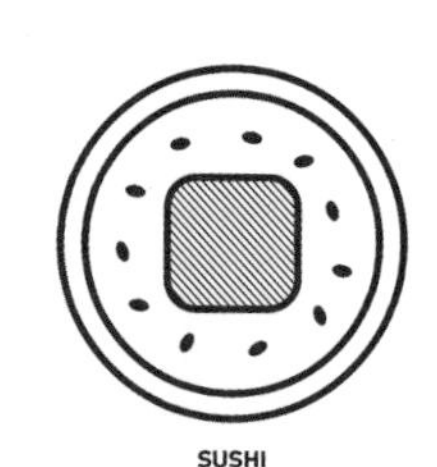
SUSHI

COCKTAIL

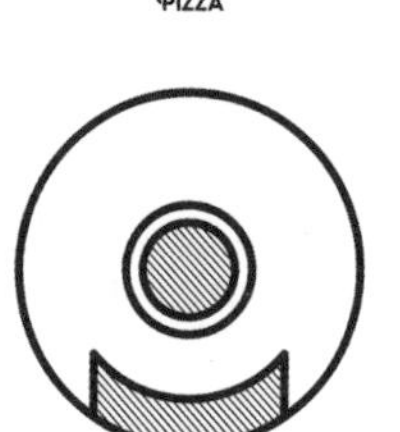
WINE

BAKERY

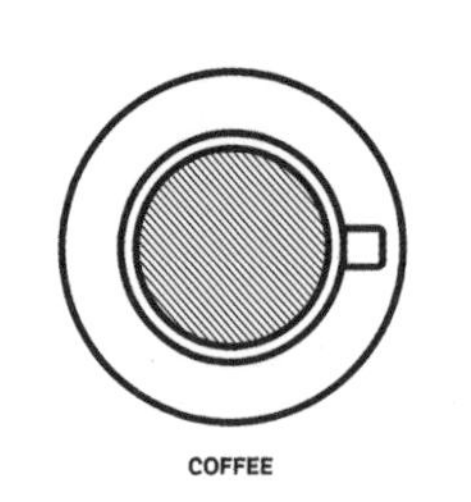
COFFEE

LISBON, PORTUGAL

ESTE OESTE

Design Expression of Cultural Variety

东西餐厅 / 葡萄牙，里斯本

多样性文化的设计体现

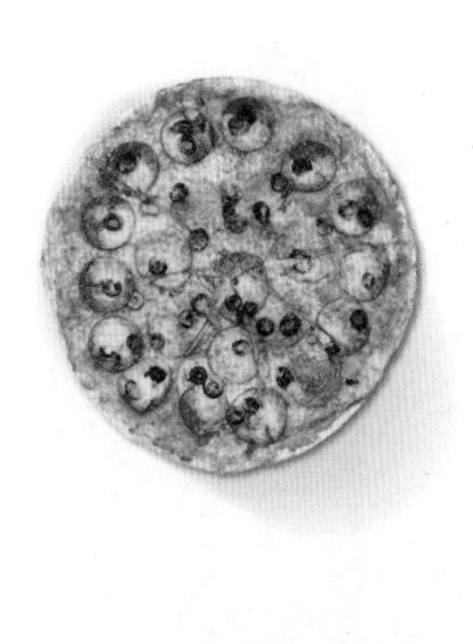

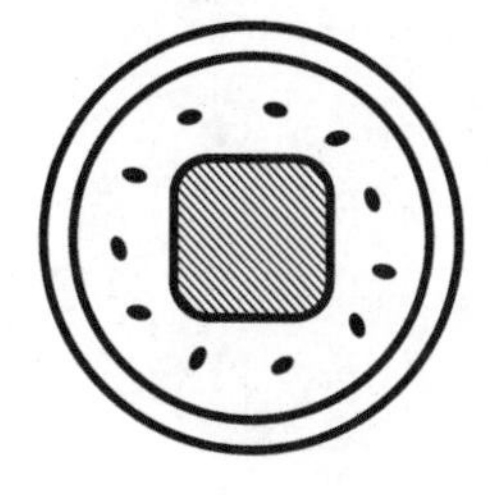

GOMA was asked to create the name and visual identity for a restaurant with one particularity: it has two cuisines, Italian and Japanese. It is located in Belém, the heart of Lisbon's tourist route and the starting point of the "Descobrimentos" so the designers also wanted to communicate the cultural diversity of the people who visit this place and of the place itself. And so Este Oeste (east west) was born.

Design agency: GOMA Designer: Diana Sousa Photography: André Carvalho Client: Este Oeste

GOMA 设计公司受邀为一家餐厅命名并打造视觉形象设计。这家餐厅供应两种美食：意大利菜和日本料理。餐厅位于里斯本热门旅游线路的中心——"发现者纪念碑"的起点，因此设计师决定在设计中表现游客和该地点本身的文化多样性。这也是餐厅的名字"东西"的来源，它代表着东西方的融合。

设计机构：GOMA 设计公司 设计师：戴安娜·苏萨 摄影：安德烈·卡瓦略 委托方：东西餐厅

JAPANESE RESTAURANT

日料店

TOKYO, JAPAN

50%, TRANSLUCENT RESTAURANT

A Fresh and Aesthetic Restaurant in Tokyo

50% 透明餐厅 / 日本，东京

“人间有味是清欢”，来自东京的清新餐厅

How would be the branding of a semitransparent restaurant? All the dishes selected and cooked in this restaurant have the characteristic to be translucent and the designers wanted to keep this visual peculiarity for all the physical extents, from bags to menu. So they've played with soft colours and translucent papers.

Designers: Matteo Morelli, Yurika Omoto, Tomomi Kuniki, Chinae Takedomi, Asuka Miyakoshi, Sayoko Aoki Photography: Matteo Morelli

B E H
I N D
50%
The stories, the faces, the names behind the translucent restaurant.
半透明レストランの物語

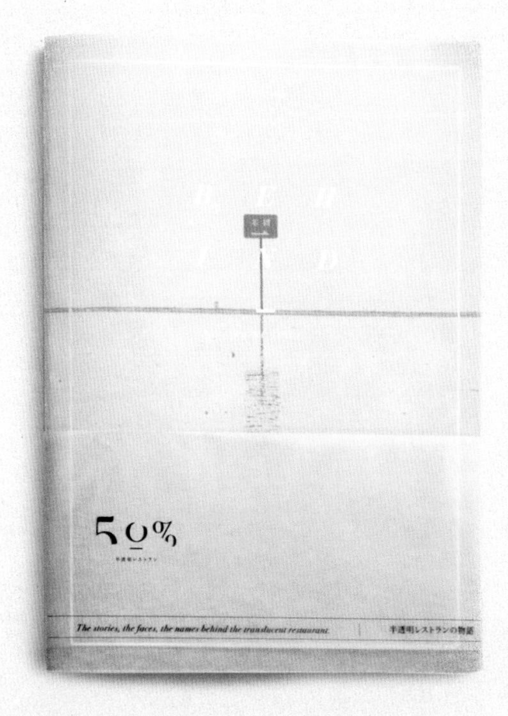
50%
The stories, the faces, the names behind the translucent restaurant.
半透明レストランの物語

50%
The stories, the faces, the names behind the translucent restaurant.
半透明レストランの物語

B E H
I N D
AUTUMN ISSUE
50%
The stories, the faces, the names behind the translucent restaurant.
半透明レストランの物語

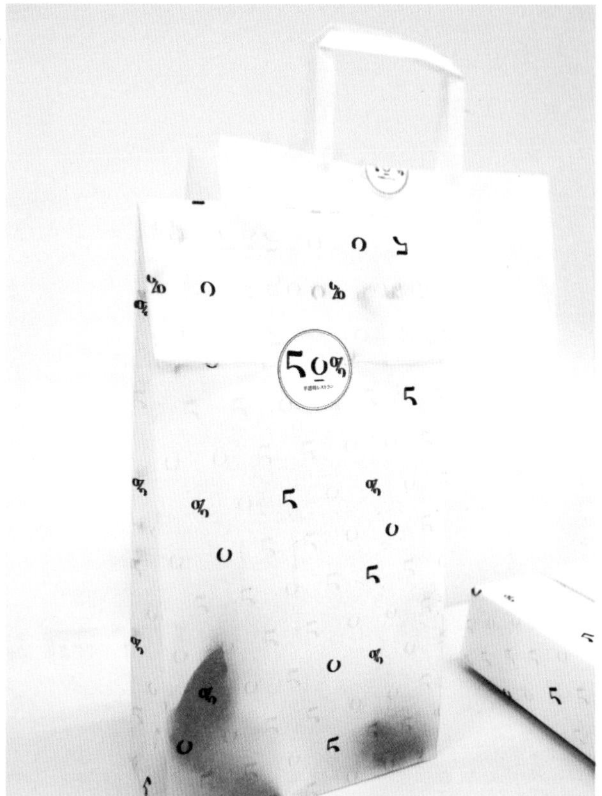

何为半透明的餐厅进行品牌设计呢？这家餐厅所有的菜品都具有半透的特色，设计师想将这一特色延续到纸袋、菜单等一系列实物中，因他们选择了柔和的色彩和半透明的纸张。

计师：马泰奥·莫雷利、大本由里香、国木智美、武富新奈、宫腰明香、青木小夜子 摄影：马泰奥·莫雷利

SINGAPORE

TANUKI RAW

The Tanuki and Raw Industrial Style

狸猫原生餐厅 / 新加坡

一只狸猫与“原生”工业风格

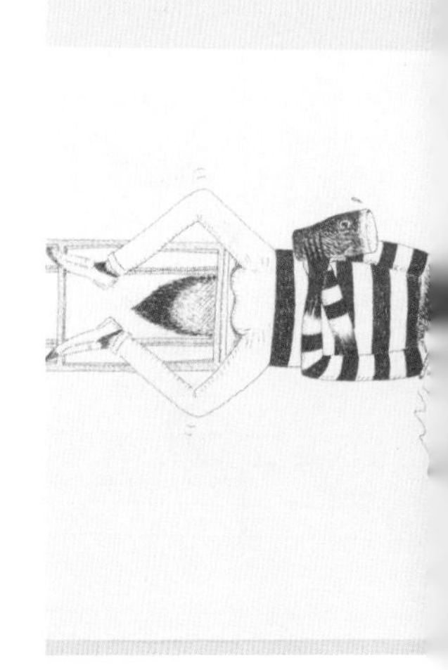

A tanuki is a Japanese racoon dog often seen as an auspicious figurine welcoming guests, so it became the natural mascot for Tanuki RAW, a Japanese martini/sushi bar. The designers kept the brand identity fairly minimal and to keep to the raw theme, designed collaterals that were letter pressed onto grey boards. The interiors were completely gutted out for a stripped down industrial look, complete with a massive concrete bar set in the middle of the space.

Design agency: Bureau Creative director: Kai Yeo Designers: Kai Yeo, Anthony Lew Illustrator: Loo Lay Hua Interior / Furniture consultant: George B.K. Soo, FLIQ Client: Tanuki Raw

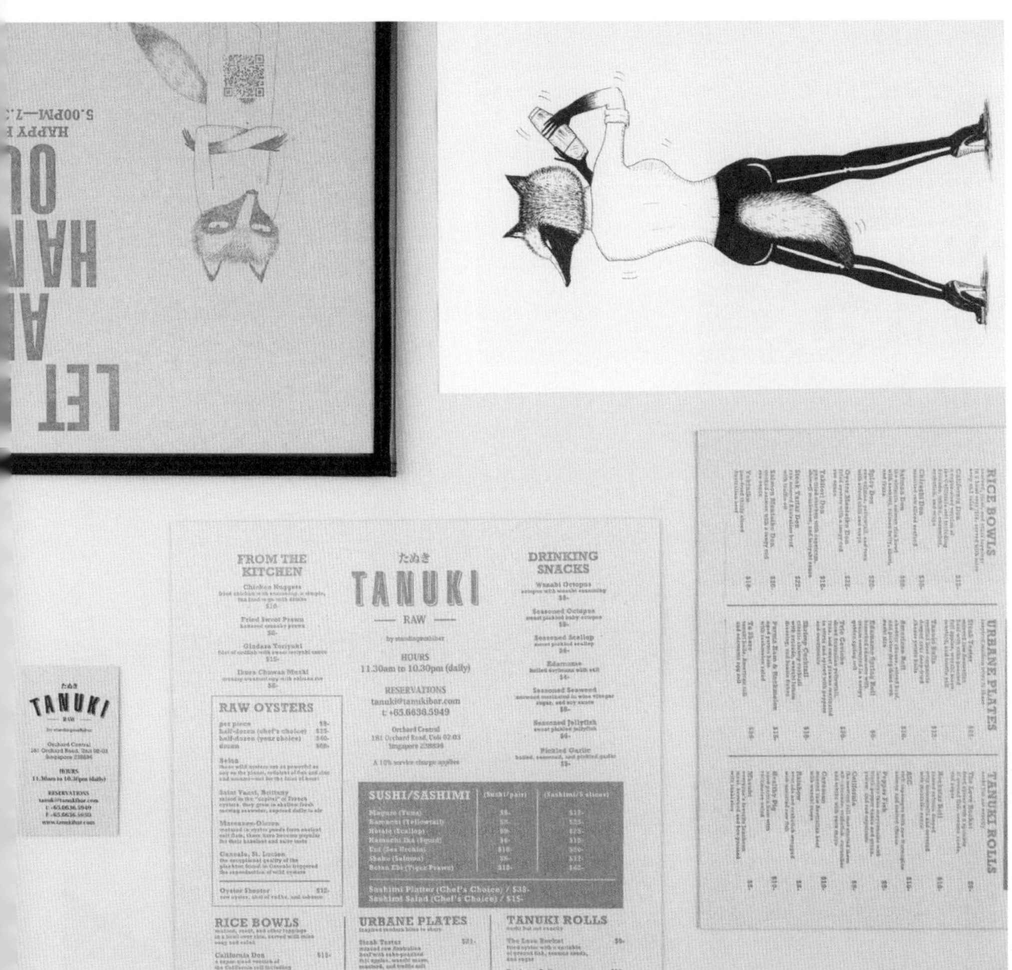

狸猫在日本文化中通常被看作是喜迎宾客的吉祥物，所以作为一家日本料理餐厅，狸猫被理所当然的作为了吉祥物。设计师将品牌标识设计的十分简洁，以对应“原生”的主题。菜单等纸制品全部采用铅字打印在灰色纸板上。室内设计完全采用粗糙的工业风格，空间正中有一个大型混凝土吧台。

设计机构：BUREAU ALLS 设计公司 设计总监：卡伊·约 设计师：卡伊·约、安东尼·刘 插画师：卢雷华 室内 / 家具顾问：乔治·B.K·孙、FLIQ 公司 委托方：狸猫原生餐厅

BALTIC SEA

SUSHI & CO.

A Combination of Scandinavian Elements and Oceanic Symbols

寿司公司 / 波罗的海

斯堪的纳维亚元素与海洋元素的交融

Sushi & Co. is a sushi restaurant on a Baltic Sea cruise ship that was in the need of a new visual identity. Bond designed a simple and clever logo and a brand identity. The new design incorporates Scandinavian elements with a sophisticated colour scheme and oceanic symbols.

Design agency: Bond Creative Agency Designer: Toni Hurme Photography: Angel Gil Client: Tallink Silja Line

23
34
38
Sushi & Co.
RESTAURANT & CAFE
BREAKFAST, LUNCH & DINNER
DRINK LIST

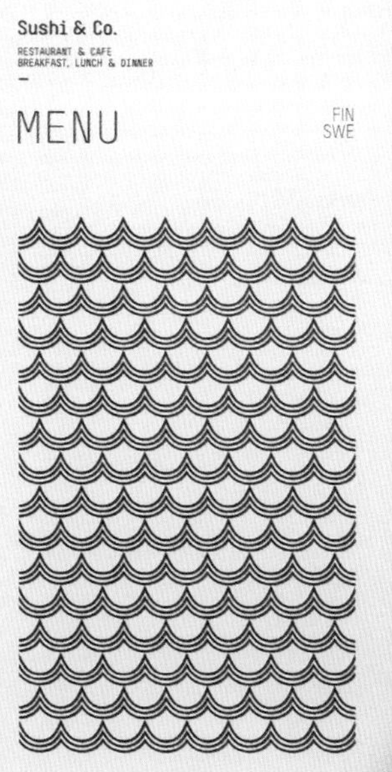
Sushi & Co.
RESTAURANT & CAFE
BREAKFAST, LUNCH & DINNER
MENU
FIN
SWE

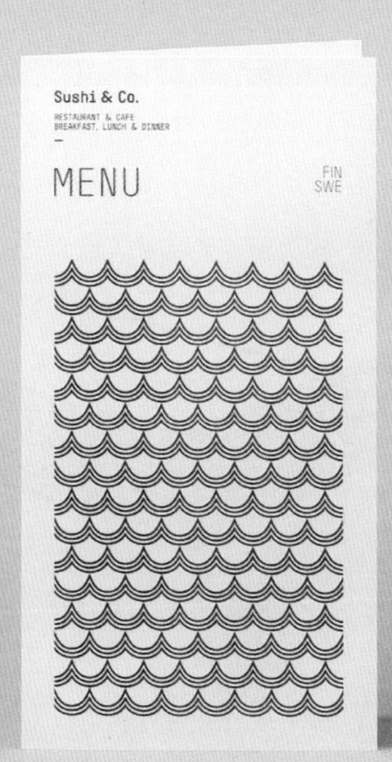
Sushi & Co.
RESTAURANT & CAFE
BREAKFAST, LUNCH & DINNER
MENU
FIN
SWE

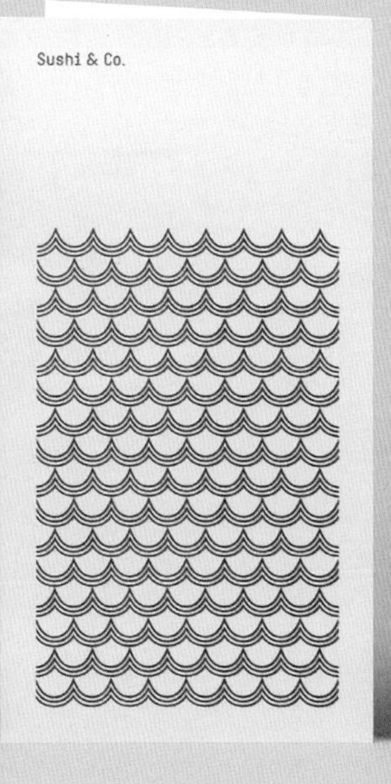
Sushi & Co.

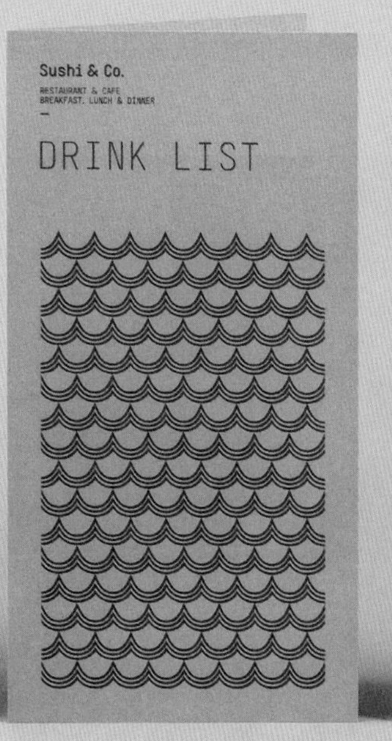
Sushi & Co.
RESTAURANT & CAFE
BREAKFAST, LUNCH & DINNER
DRINK LIST

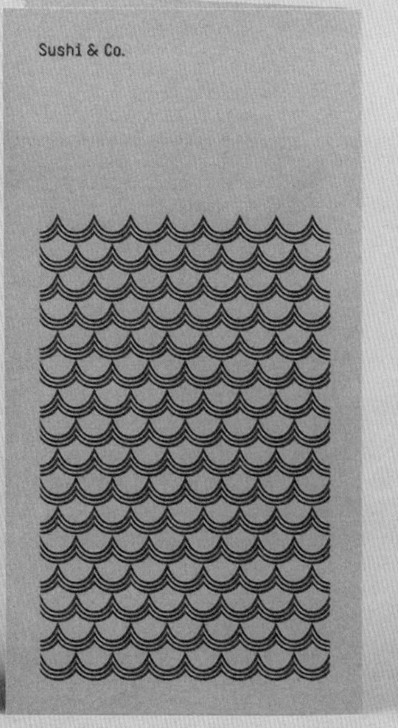
Sushi & Co.

Sushi & Co.
MAKI NIGIRI SASHIMI
Combo
SMALL
MEDIUM
LARGE
Maki
KAPPA
SHAKE MAKI
SPICY SHRIMP
CALIFORNIA ROLL

MAKI NIGIRI SASHIMI
Platters
SPECIAL
SASHIMI CHIRASHI
"SURF & TURF"
VEGETARIAN PLATTER
SALMON SUSHI
Side orders
GOMAWAKAME
MISO

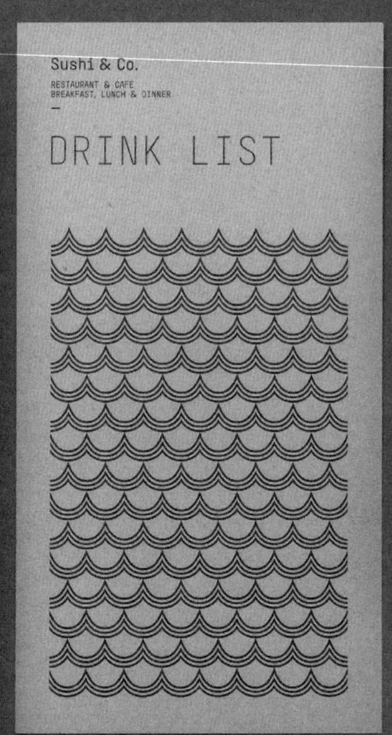
Sushi & Co.
RESTAURANT & CAFE
BREAKFAST, LUNCH & DINNER
DRINK LIST

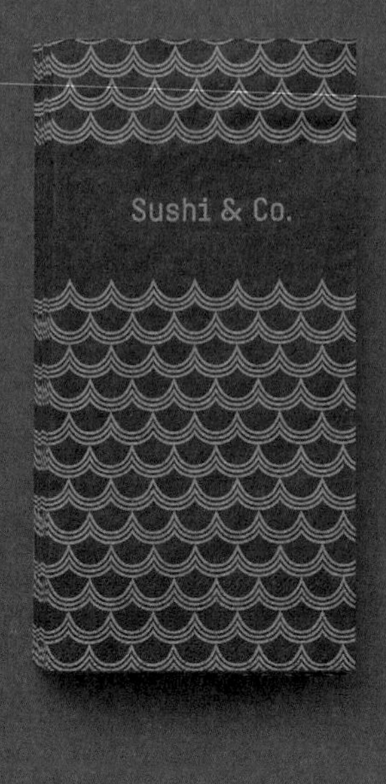
Sushi & Co.

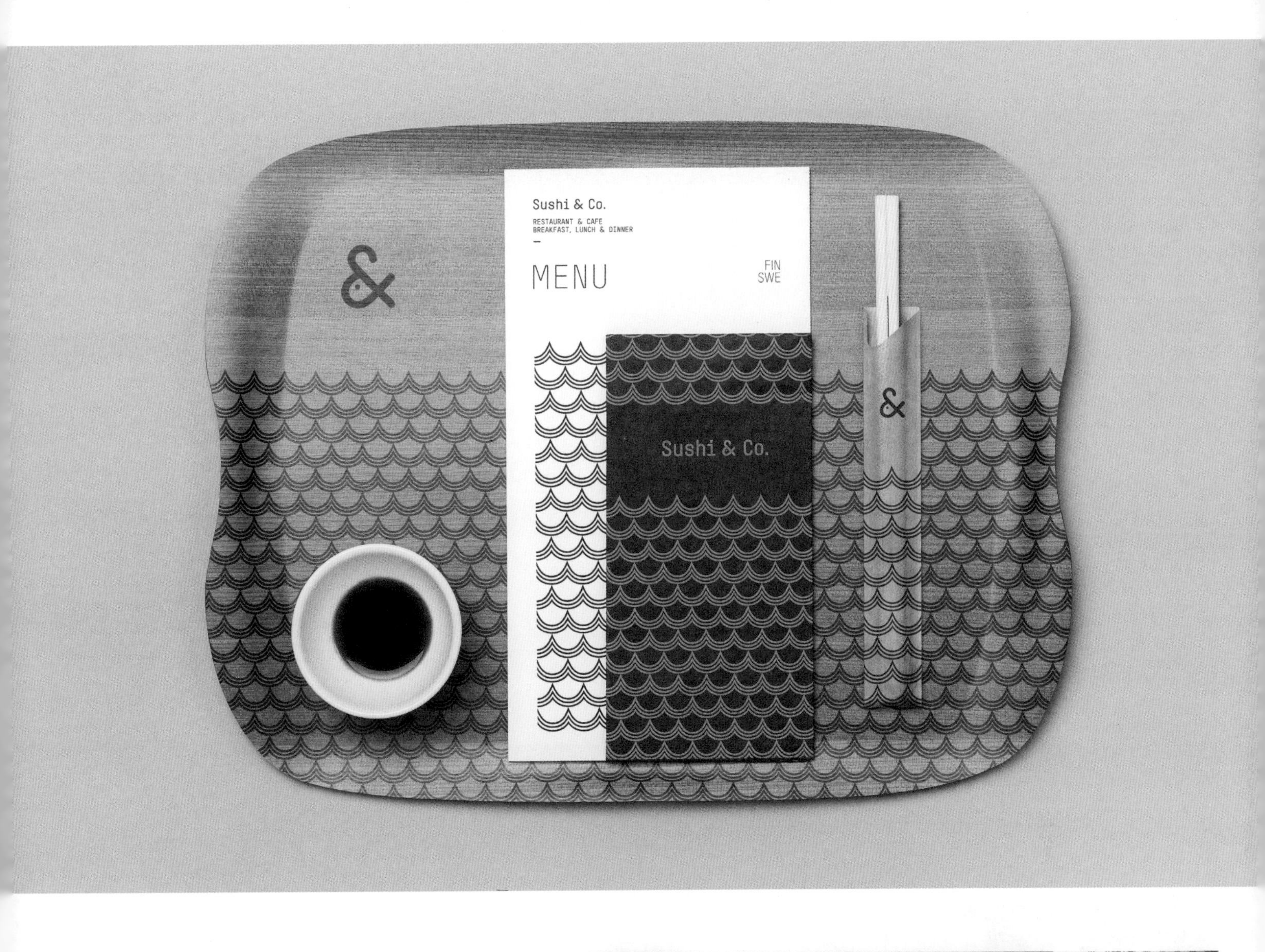

寿司公司是一家位于波罗的海游轮上的寿司餐厅，餐厅需要打造一套全新的视觉形象。Anagrama Bond 创意公司为其设计了一个简单而巧妙的LOGO和品牌形象。新设计融合了具有精致色彩的斯堪的纳维亚元素和各种海洋元素。

设计机构：Bond 创意公司 设计师：托尼·赫尔姆 摄影：安吉尔·吉尔 委托方：Tallink Silja 号游轮

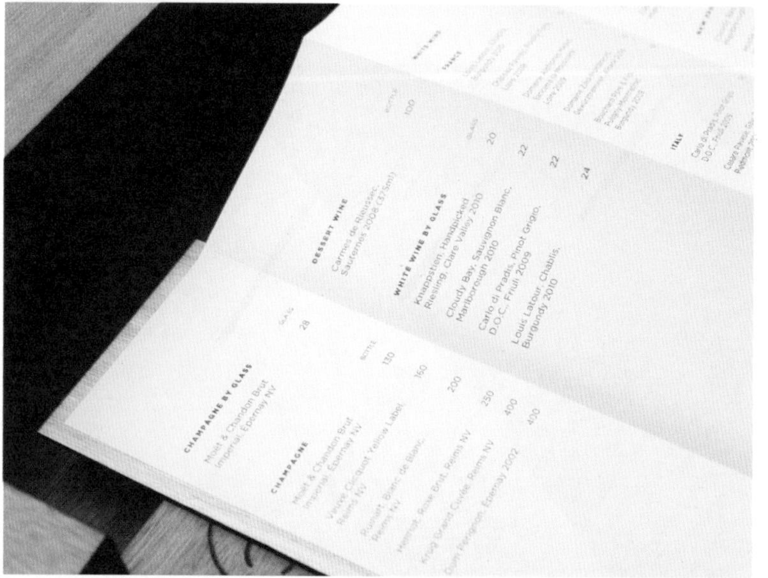

Fat Cow is a specialist beef restaurant employing the Japanese way of picking cooking and serving beef. Drawing inspiration largely from the Japanes aesthetic – Wabi Sabi with traits that include simplicity, economy, austerity, modest and the appreciation of the ingenuous integrity of natural objects and processes wood is used primarily as the platform of this brand communication. The nor uniformity and texture suggests the Wabi Sabi beauty of imperfection. The mark an the searing on the wood are also reminiscent of the branding of cattle.

Design agency: Foreign Policy Design Group Creative director: Yah-Leng Yu Designers Gwen Chan & Bryan Angelo Lim Printer: Allegro Print

肥牛餐厅是一家日式牛肉专门餐厅，以日本料理的方式挑选、烹饪和上菜。设计的灵感主要来自于日本美学的“残缺之美”（简洁、节约、朴素、谦逊以及对自然物和自然进程的欣赏。餐厅的设计主要以木材作为品牌传播的平台。质地不均的纹理正好诠释了残缺之美。木头上的印记和灼痕也应对了品牌的形象象征——牛。

设计机构：Foreign Policy 设计集团 创意总监：于亚玲 设计师：陈格温、布莱恩·安格鲁·李 印刷：Allegro 印刷公司

SINGAPORE

FAT COW

The Japanese Aesthetic – Wabi Sabi

肥牛餐厅 / 新加坡

日本美学的“残缺之美”

BEIJING, CHINA

MIU CREATIVE CUISINE

Wonderful "米" Shape Graphics

MIU 秘团 / 中国，北京

奇妙的米字图形

MIU is based on creative rice dishes, complemented by specialty coffee, tea and delicate desserts. MIU is a vibrant and creative urban fashionable food and beverage brands. The designers take the shape of the word " 米 " to create the guidelines in order to create more " 米 " shape graphics, getting the result of a flexible and changeable logo.

Design agency: One&One Design Creative director: Li Wen Designer: Li Wen Client: MIU Creative Cuisine

MIU 秘团以创意米团为主体餐点，辅以特色咖啡茶饮、精致甜品，是一个充满活力和创造力的都市时尚餐饮品牌。设计师以米字格为主体依据，创作出不同的米字图形，从而得到一个灵活、具有多变性的 LOGO 形态。

设计机构: One&One 设计公司 创意总监: 温力 设计师: 温力 委托方: MIU 秘团餐饮店（北京）

PARIS, FRANCE

NAGOYA

Natural Beauty of Oriental Zen

名古屋日本料理 / 法国，巴黎

东方禅意的自然之美

Nagoya Japanese Cuisine is a Japanese cuisine brand located in Pairs, France. During a decade-long cooperation, One&One designed the visual promotion system with the natural beauty of Oriental Zen all along. Japanese Cuisine also known as 'washoku', which stand for natural and original are the main spirits of Japanese cuisine. The concept of the Japanese Government apply 'washoku' to be world heritage mentioned: 'Washoku' embodied the Japanese spirit 'respect for nature'; It reflects the beauty of nature and the changes of the four seasons; Japanese cuisine keeps the food fresh and original and also focuses on multiplicity....

Design agency: One&One Design Creative director: Li Wen Designer: Li Wen Client: Nagoya Japanese Cuisine

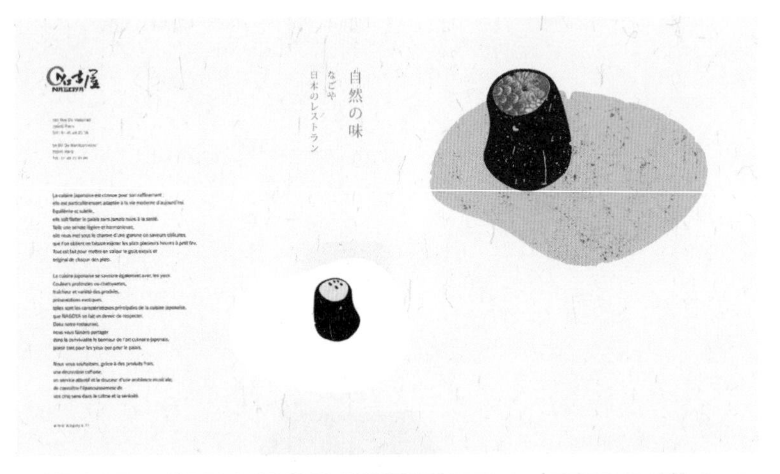

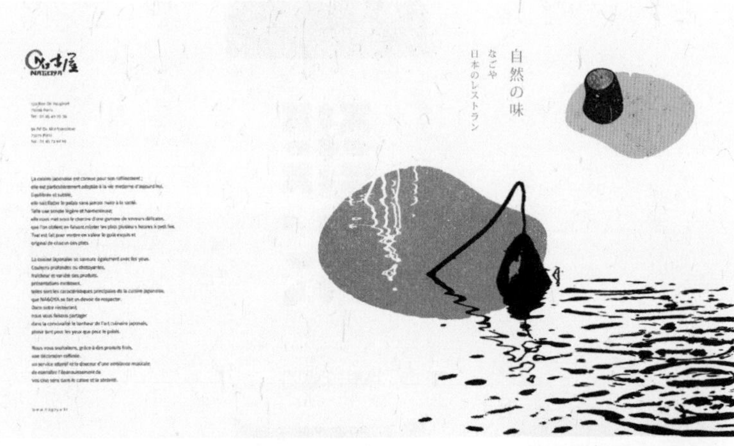

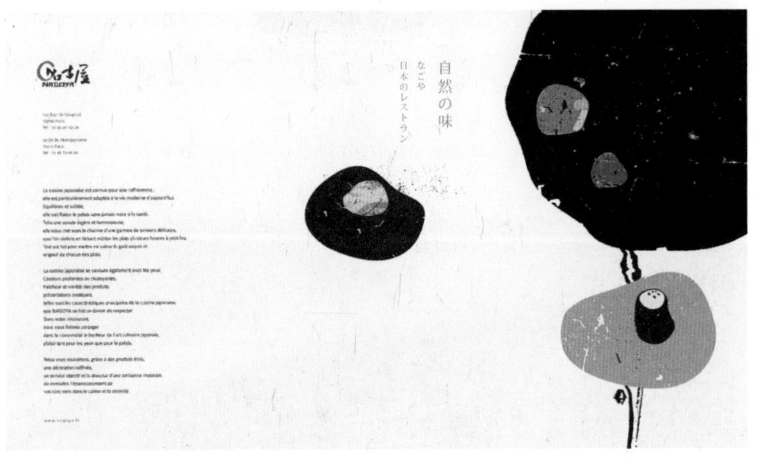

名古屋日本料理是位于法国巴黎的一个日本料理品牌，在长达十年之久的合作中，设计师始终以东方禅意的自然之美为料理店做视觉推广。日本料理也称为“和食”，自然原味是日本料理的主要精神，在日本政府提出“和食”申遗的理念中提到：“和食”体现了日本人“尊重自然”的精神；体现了自然之美以及节气的变化；日本料理保持食材的新鲜和原味，注重视觉和味觉的多样性……

设计机构：One&One 设计公司 创意总监：温力 设计师：温力 委托方：NAGOYA 名古屋日本料理店（巴黎）

鮮
なごや
日本のレストラン
RESTAURANT JAPONAIS
FONDÉ EN 1996
NAGOYA

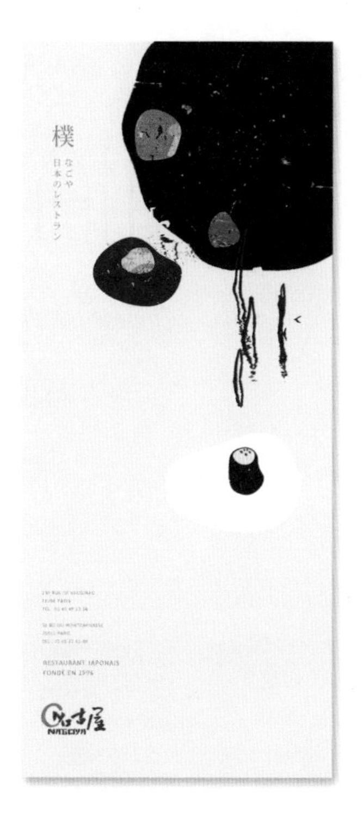
樸
なごや
日本のレストラン
RESTAURANT JAPONAIS
FONDÉ EN 1996

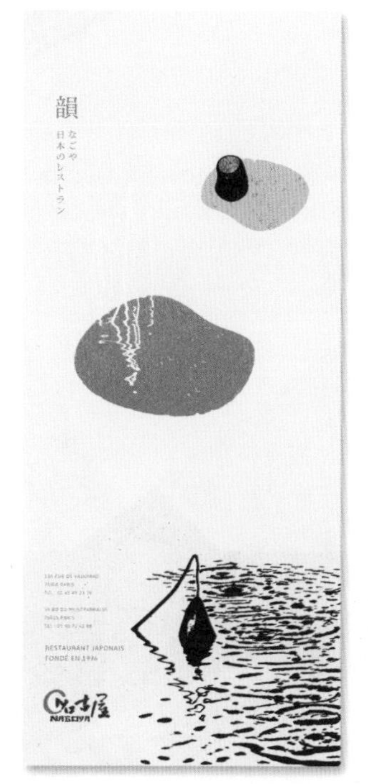
韻
なごや
日本のレストラン
RESTAURANT JAPONAIS
FONDÉ EN 1996
NAGOYA

真
なごや
日本のレストラン
RESTAURANT JAPONAIS
FONDÉ EN 1996

自然の味
なごや
日本のレストラン
NAGOYA
RESTAURANT JAPONAIS
FONDÉ EN 1996

自然の味
なごや
日本のレストラン
NAGOYA
RESTAURANT JAPONAIS
FONDÉ EN 1996

NAGOYA
自然の味
なごや
日本のレストラン
High Tea
15€
Cash Coupon

PARIS, FRANCE

MATSUYAMA

Chinese Ink Painting as An Exotic Symbol in the West

松山日本料理 / 法国，巴黎

水墨元素在西方的异域象征

MATSUYAMA is a brand of Japanese Cuisine in Paris, france. One&One Design uses Chinese ink painting with eastern artistic conception as its core visual element to reflect the emotional appeal of the foreign lands that the eastern cooking culture spreads in the west.

Design agency: One&One Design Creative director: Li Wen Designer: Li Wen Client: Matsuyama Restaurant Japan

MATSUYAMA 松山是位于法国巴黎的一个日本料理品牌，设计师用东方意境的水墨作为其核心的视觉元素，体现东方饮食文化在西方散发的异域情调。

设计机构：One&One 设计公司 创意总监：温力
设计师：温力 委托方：松山日本料理店（巴黎）

VALENCIA, SPAIN

NOZOMI SUSHI BAR

"Emotional Classic" and "Rational Contemporary"

希望寿司 / 西班牙，巴伦西亚

"感性的古典"与"理性的现代"的双重性

The project in which Masquespacio began to work in January 2014 starts with a previous study of Japanese culture and the origin of sushi. A study in which was involved the whole team of the Spanish creative consultancy to understand and represent the Japanese culture through the brand image and specially through the interior of the new restaurant from José Miguel Herrera and Nuria Morell. The brand name Nozomi was chosen by the founders of this project being a 'Japanese high speed bullet train' and at the same time meaning 'fulfilled dream'; two significances with which with José Miguel and Nuria felt identified and that create a duality present continuously through the whole project: "Emotional classic' and 'Rational contemporary'. Starting with the brand image we can see how this duality is represented on one way as 'Rational contemporary' through the Western typography, while on the other the hiragana (Japanese writing) shows the "Emotional classic' touch through its logo. In the meantime the cherry-tree's flowers, inspired by the origami, bloom naturally.

Design agency: Masquespacio Creative director: Ana Milena Hernández Palacios Photography: David Rodríguez and Carlos Huecas Client: Nozomi Sushi Bar

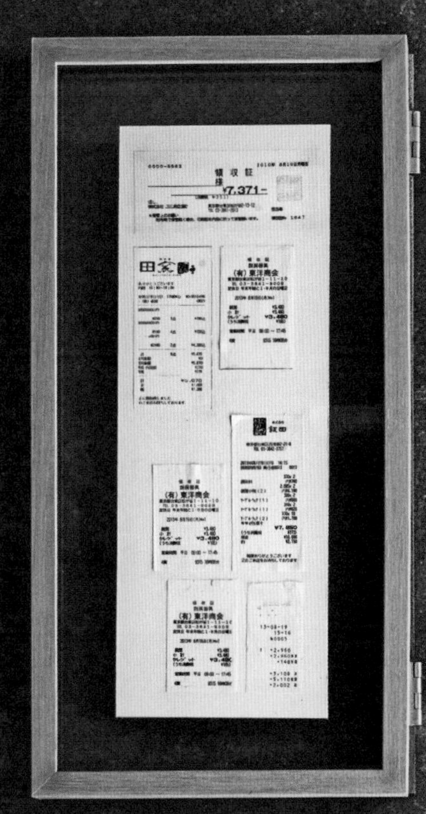

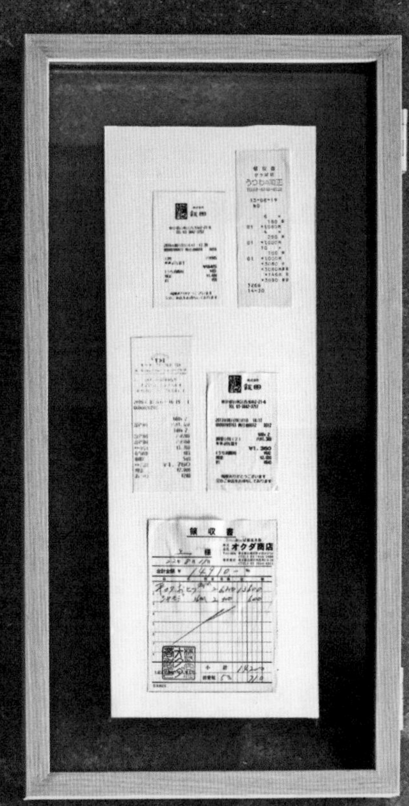

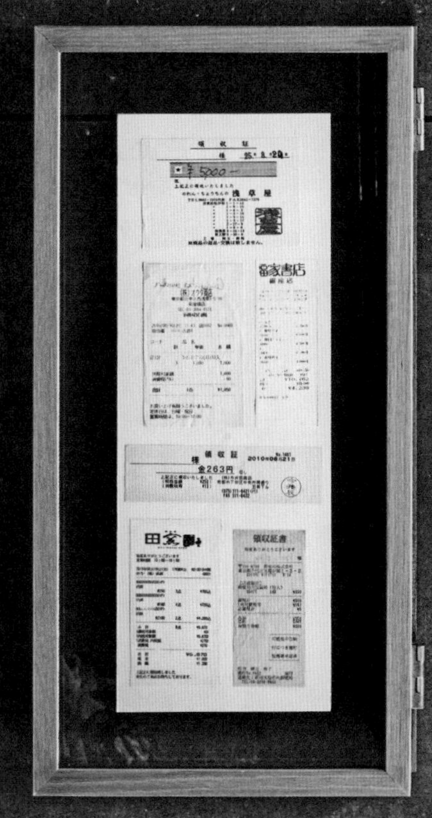

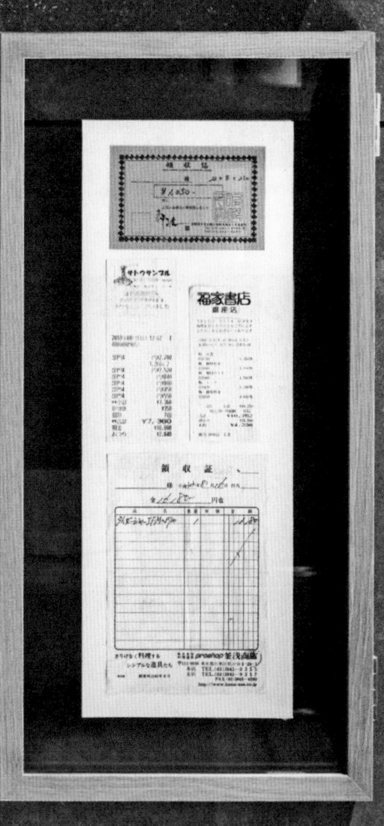

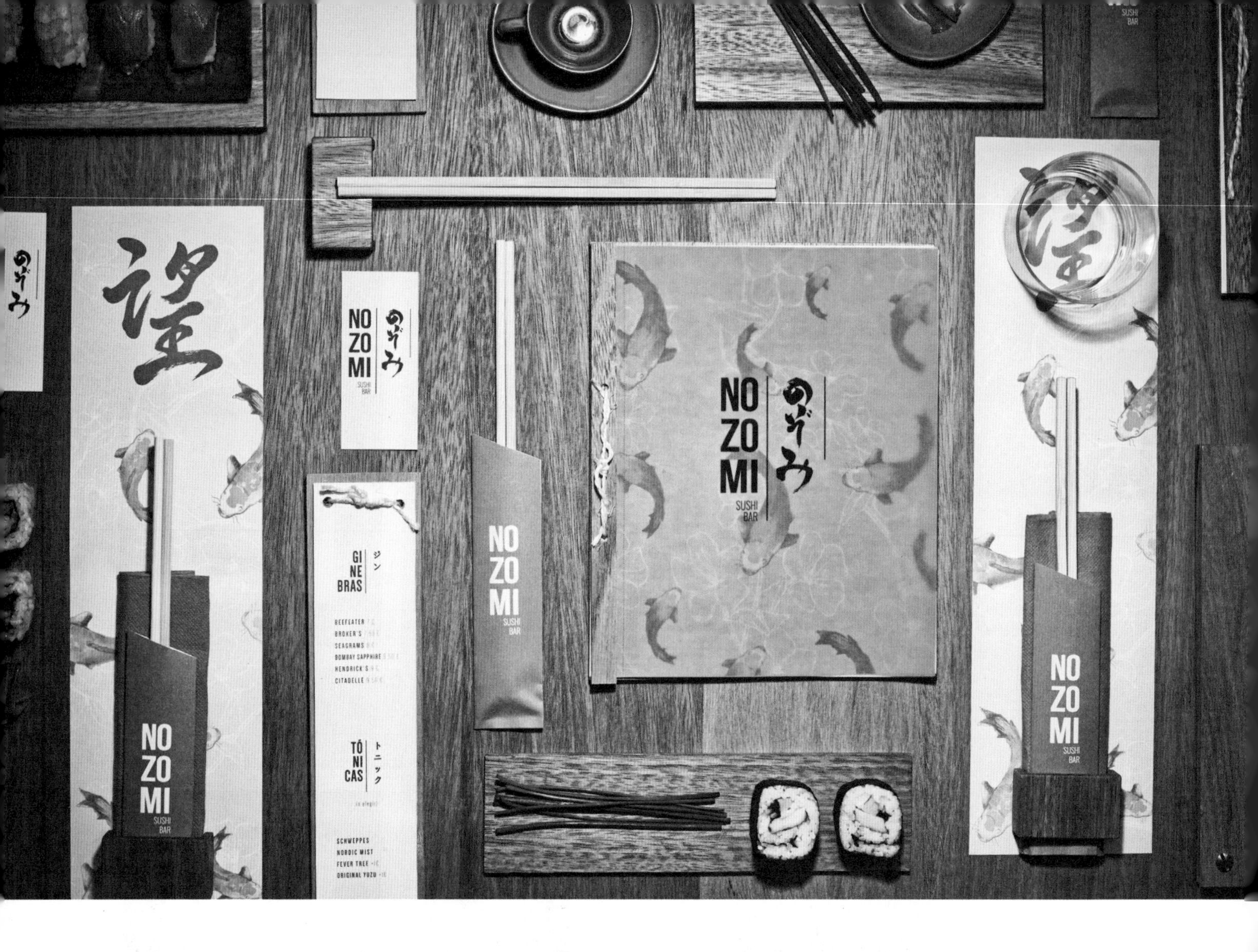
NO
ZO
MI
のぞみ
SUSHI BAR
GI
NE
BRAS
ジン
BEEFEATER
BROKER'S
SEAGRAMS
BOMBAY SAPPHIRE
HENDRICK'S
CITADELLE
TÓ
NI
CAS
トニック
SCHWEPPES
NORDIC MIST
FEVER TREE
ORIGINAL YUZO

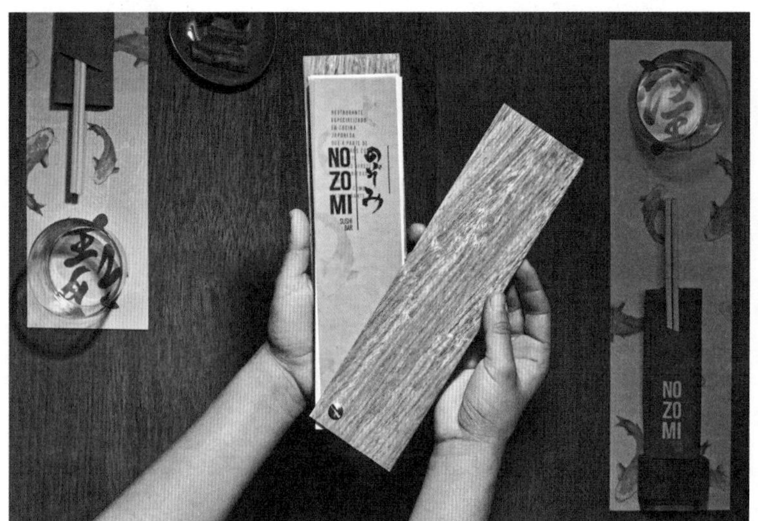
NO
ZO
MI
のぞみ
SUSHI BAR
NO
ZO
MI

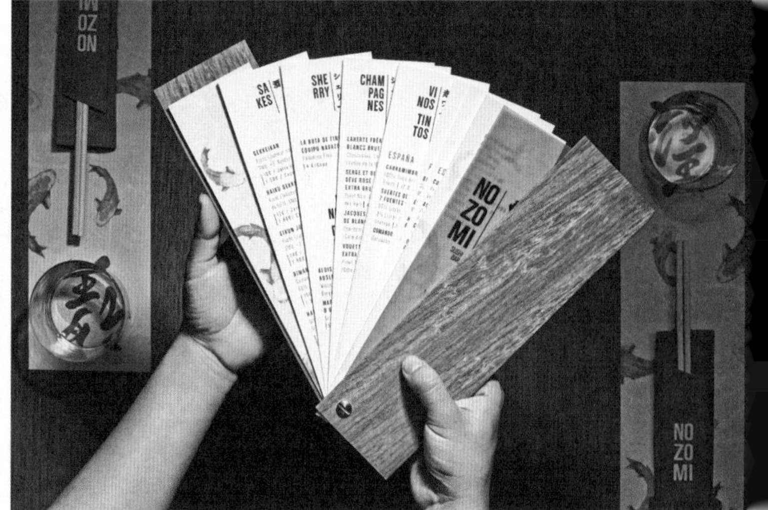
SA
KES
SHE
RRY
CHAM
PAG
NES
VI
NOS
TIN
TOS
ESPAÑA
NO
ZO
MI
NO
ZO
MI

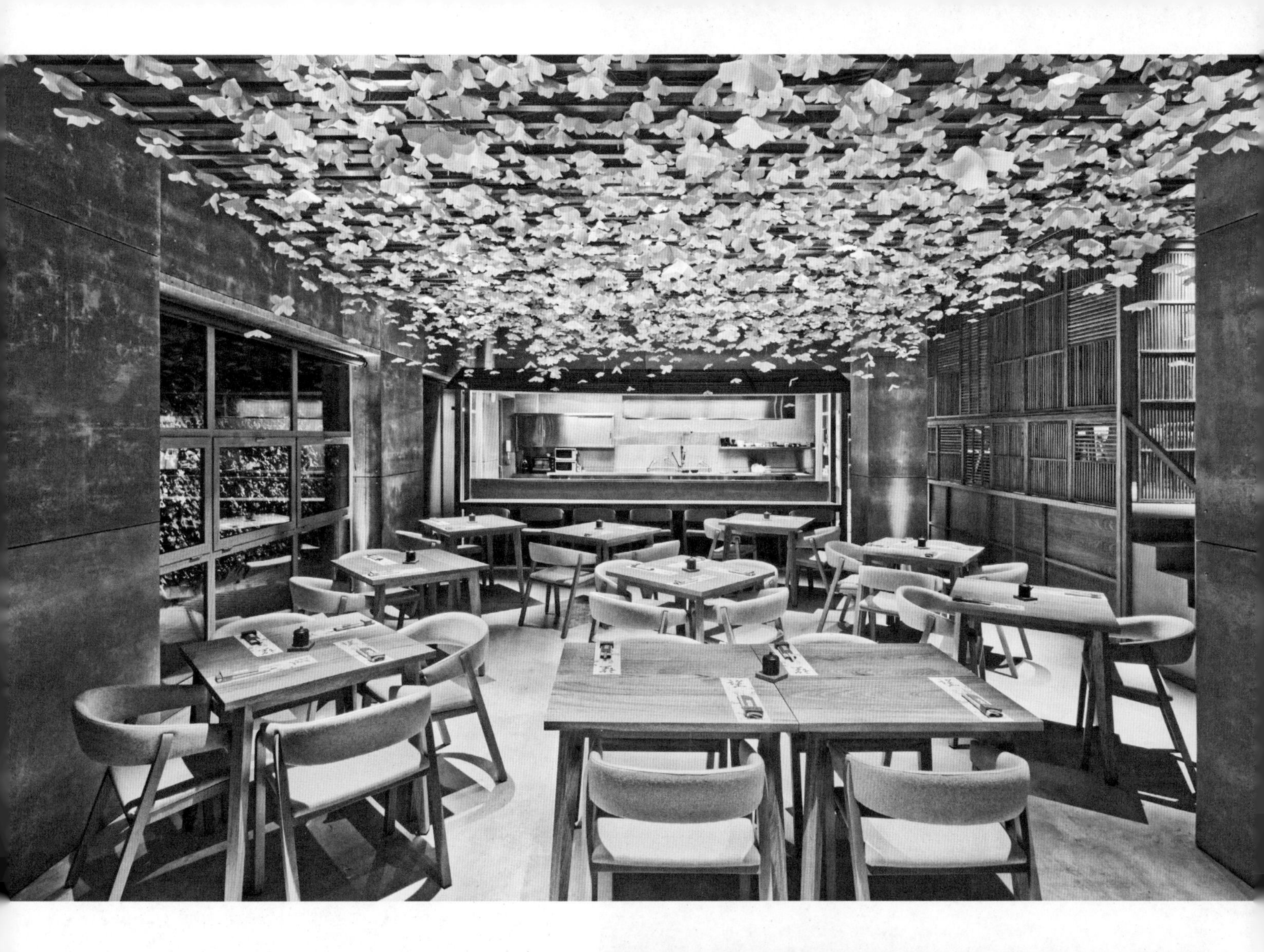

Masquespacio 于 2014 年 1 月开始着手设计这个项目，他们首先对日本文化和寿司的起源进行了调研。这家西班牙创意咨询公司的全体成员都参与了调研，力求通过餐厅的品牌形象和室内设计表现日本文化。餐厅的名称“希望”由创始人所选，既代表着日本高速列车“希望号”，又意味着“梦想的实现”。这两种意义定义了整个项目的双重性：“感性的古典”和“理性的现代”。餐厅的品牌形象通过西方字体呈现了“理性的现代”；另一方面，日本平假名的应用则通过 LOGO 展示了“感性的古典”。在天花板上，象征日本文化的折纸樱花自然的绽放。

设计机构：Masquespacio 设计公司 创意总监：安娜 · 米莱娜 · 赫尔南德 · 帕拉西奥斯 摄影：大卫 · 罗德里格斯、卡洛斯 · 韦卡斯 委托方：希望寿司

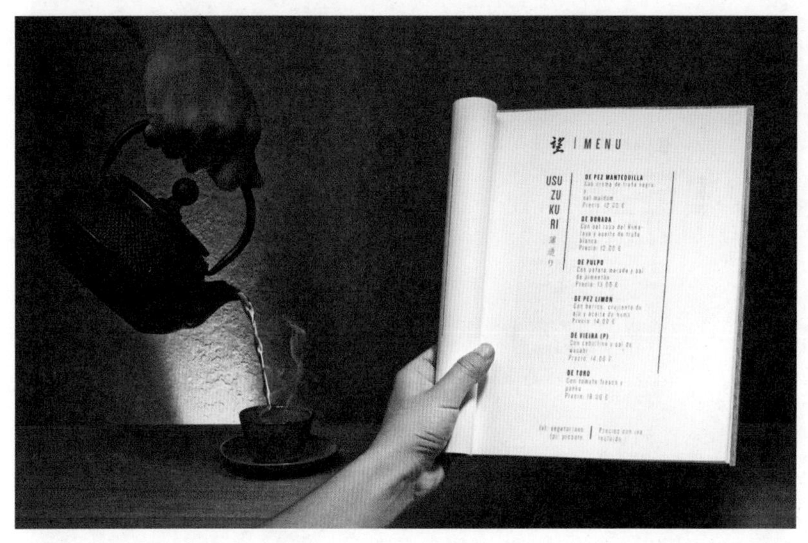

MONTERREY, MEXICO

TORO TORO

Visual Environment Reminiscent of a Night in Tokyo

鲔鱼餐厅 / 墨西哥，蒙特雷

东京夜生活般的视觉环境

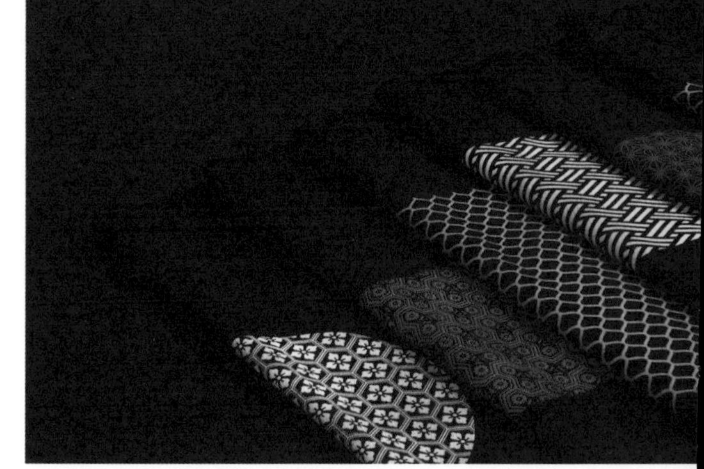

More than a Japanese restaurant, Toro Toro is a restaurant in Japan. Its visual identity is based on authenticity, and it is an exploration of the visual culture of one of the most vibrant, influential and interesting cities in the world. The design proposal was to create a visual environment that was reminiscent of a night in Tokyo, with its lights, its dynamism, its culinary delights and its culture.

Design agency: Savvy Studio Creative directors: Rafael Prieto & Raul Salazar Designer: Eduardo Hernandez Photography: Alejandro Cartagena Client: Toro Toro

鲔鱼餐厅不仅是一家日本料理餐厅，而且能让你有置身日本之感。它的视觉形象以真实为基础，探索了东京这座全球最具活力、影响力和趣味的城市之一的视觉文化。设计方案旨在通过灯光、活力、美食和文化打造一个与东京夜生活相似的视觉环境。

设计机构：Savvy 工作室 创意总监：拉斐尔·普列托、劳尔·萨拉萨尔 设计师：爱德华多·赫尔南德斯 摄影：亚力山卓·卡塔赫纳 委托方：鲔鱼餐厅

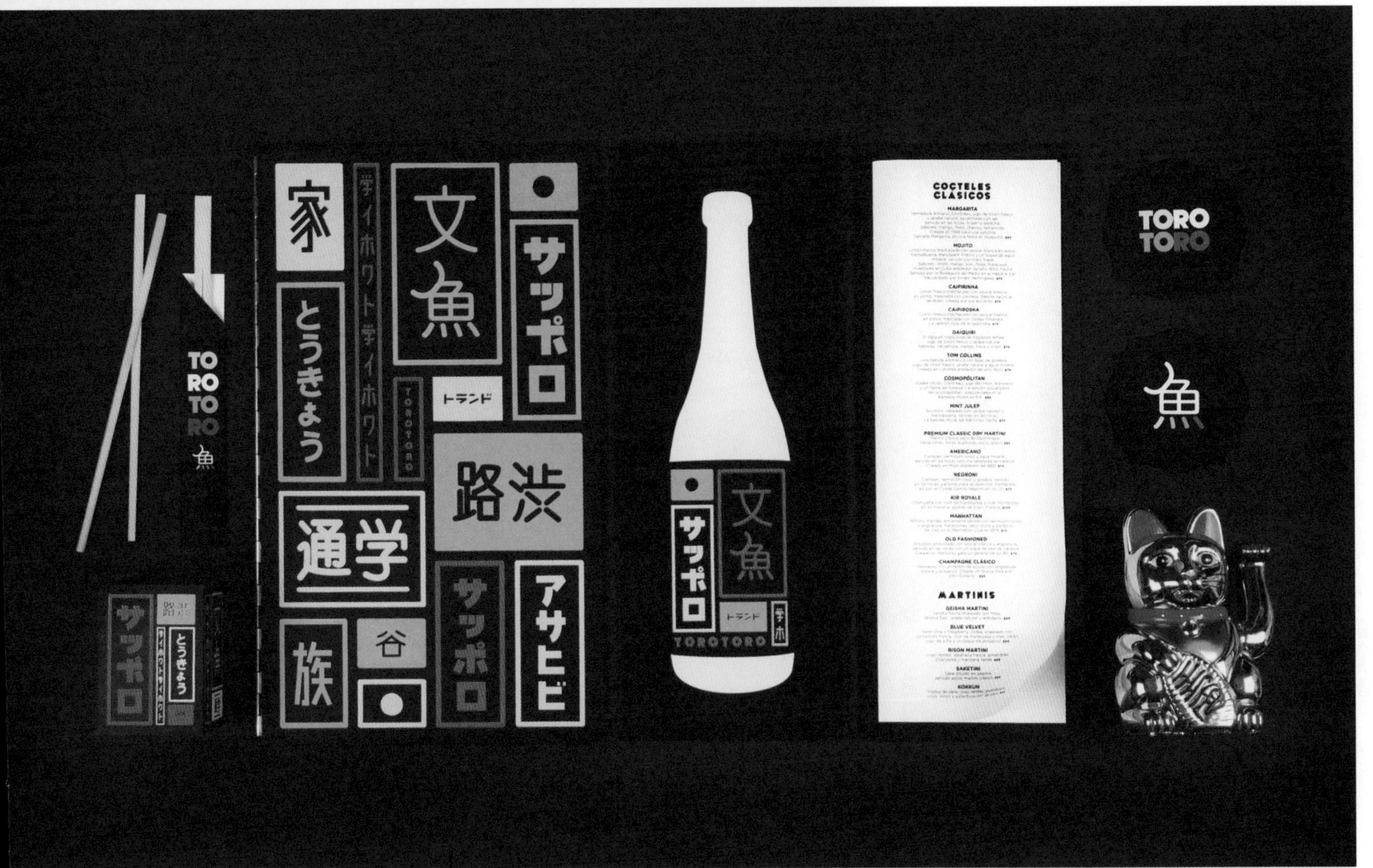

von
& zu

MEDITERRANEAN
RESTAURANT
地中海餐厅

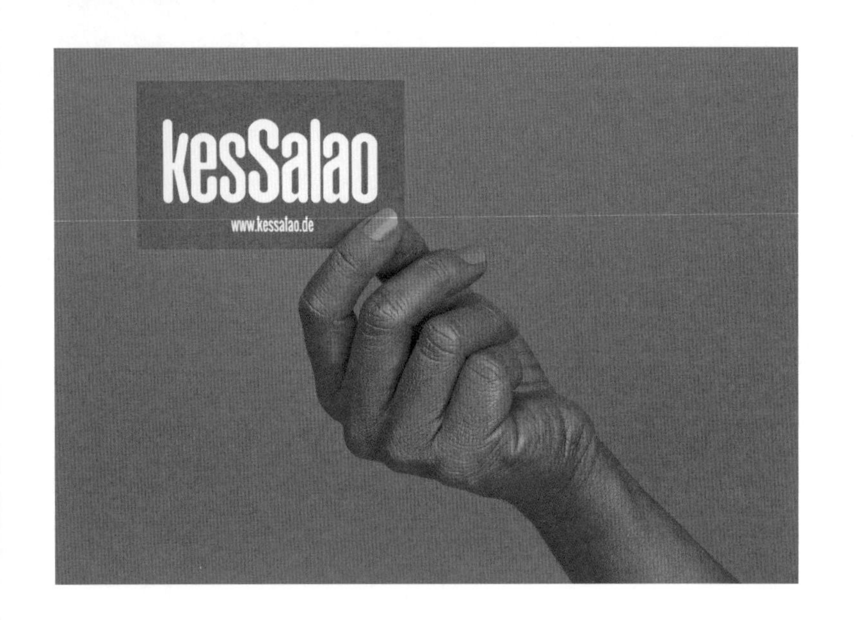

BONN, GERMANY

KESSALAO

A Colourful Restaurant from the City of Beethoven

凯斯萨劳快餐店 / 德国，波恩

来自贝多芬故乡具有糖果般色彩缤纷的餐厅

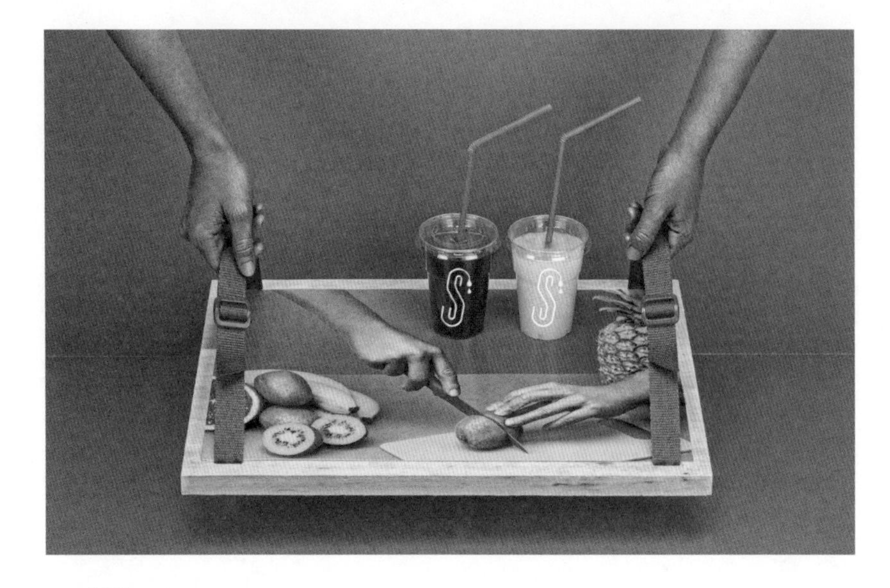

panish creative consultancy Masquespacio presents their last project realised the city of Bonn, Germany. The project consists in the brand image and erior design for Kessalao, a new take away establishment of Mediterranean od in the city of Beethoven. Everything starts from the brand image and it's ming that forms wordplay of the German "Kess" and the Spanish "Salao", th traduced as cool and amusing boy. Being a play of words in two different nguages combined by an s, a capital S needed to distinguish both words. n the other hand the brand symbol was inspired by olive oil, as the basic and ncipal product of the Mediterranean food, represented here by the drop that erprets the natural product's richness.

sign agency: Masquespacio Creative director: Ana Milena Hernández lacios Photography: David Rodríguez and Carlos Huecas Client: Kessalao

In so far as the interior design it's presented by a space that symbolises the freshness of the brands' name through a range of most popular colours for Germans. Materials like wood coming from the birch veneer used for the walls and pine for the furniture, where chosen to offer a natural look to the space. Moreover through different decorative elements made of raffia as for the seats and pots a Mediterranean touch is added repeatedly. Ana Milena Hernández Palacios: "I wanted to metaphorise the recollection of fruits and vegetables through the hampers of Raffia." The metal gratings are having a function as expositors for magazines, menu cards and pots, besides doing a job as tables in the store front of the space. Constructive details like the bars and folding tables are created as an adaptation to the necessities of the space according to the hour of minor or major traffic, being the space able to operate only as a take a way service at noon or as a small snack bar by night. A last detail is founded in the coating of raffia used for the barstools, also created by Masquespacio, that contain a department to leave jackets and handbags, exploiting maximally the reduced available space.

西班牙创意资讯公司 Masquespacio 向人们呈现了他们在德国波恩的最新设计，为一家位于贝多故乡的地中海美食外卖快餐店所提供了品牌形象和室内设计。快餐店的名字“Kessalao”是德语“Kes和西班牙语“Salao”的结合，它们都被用来形容“酷酷的、有趣的男孩子”。既然两个单词是字母“S”所连接起来的，那么“S”就应该通过大写而凸显出来。餐厅的品牌标志的设计灵感自于橄榄油——地中海地区最具代表性的食品之一。油滴的图案诠释了天然食品的丰富内涵。

餐厅的室内设计通过一系列深受德国人喜爱的色彩体现了餐厅品牌的新鲜感。墙壁的桦木贴面家具的松木为空间带来了一种自然之感。酒椰叶纤维所制成的座椅、花篮等装饰元素进一步凸了地中海风情。设计师安娜·米莱娜·赫尔南德·帕拉西奥斯说道：“我希望给人一种酒椰叶篮的感觉。”金属栅板起到了展示杂志、菜单和花篮的作用，同时也是店面前方的一个小吧台吧台、折叠桌等细节构造让餐厅可以根据客流量调节空间布置：午后是外卖餐厅，夜晚则变身小吃店。吧台高脚凳所使用的酒椰叶纤维包面同样由 Masquespacio 设计，最大限度地利用了有的空间。

设计机构：Masquespacio 设计公司（www.masquespacio.com）创意总监：安娜·米莱娜·赫尔南德·拉西奥斯 摄影：大卫·罗德里格斯、卡洛斯·韦卡斯 委托方：凯斯萨劳快餐店

DIE BOCATAS
SALATE
SMOOTHIES
TOPPINGS
ENTWERFE DEINEN SALAT SELBST...
1 SCHRITT
2 SCHRITT
3 SCHRITT
4 SCHRITT

CANTABRIA, SPAIN

LA BICICLETA

The Beauty of Geometry

自行车餐厅 / 西班牙，坎塔布里亚

几何图形之美

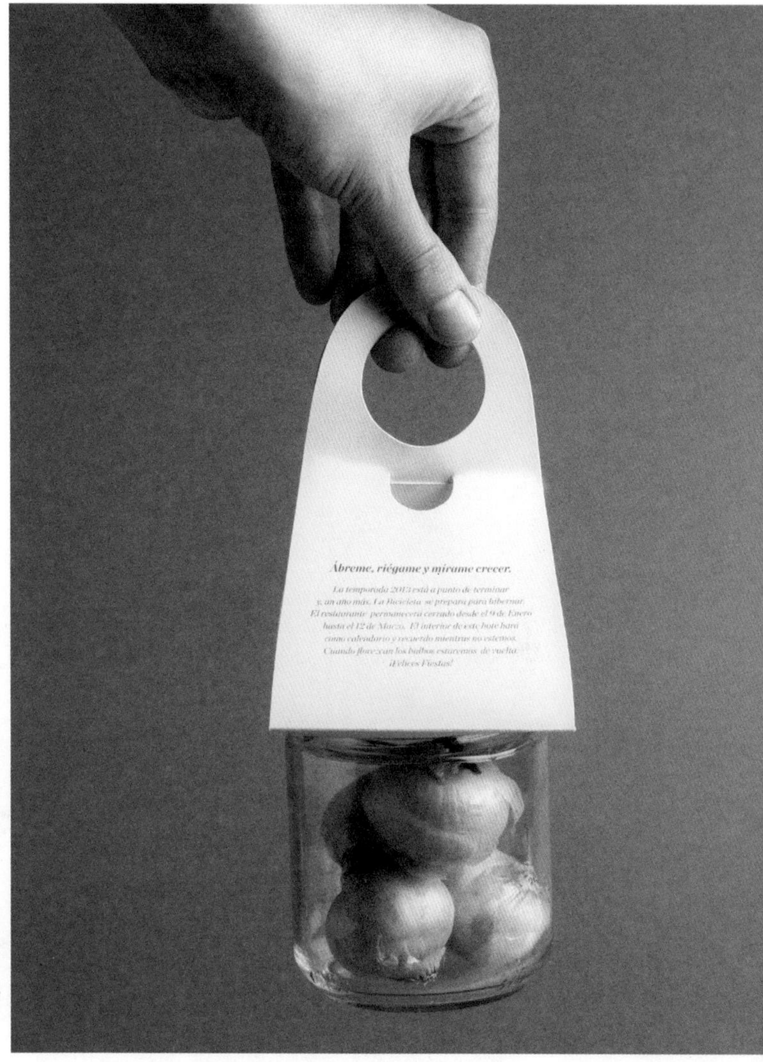

Branding design for La Bicicleta restaurant in cantabria (Spain).

Design agency: Lo Siento Studio Client: La Bicicleta

项目是为西班牙坎塔布里亚的自行车餐厅所提供的品牌形象设计。

设计机构：Lo Siento 工作室 委托方：自行车餐厅

BARBA
sea food / street food

SEAFOOD RESTAURANT
海鲜餐厅

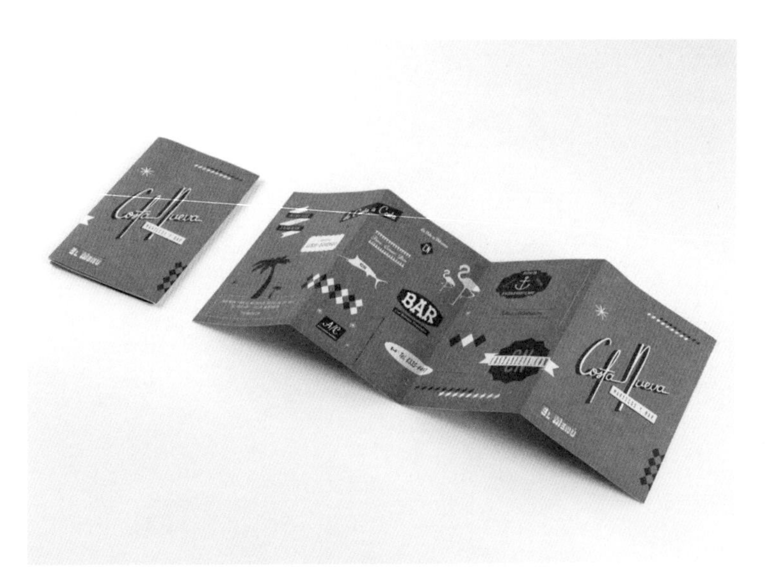

Analysis – Costa Nueva is a seafood restaurant founded by two friends obsessed with offering the city the freshest ingredients brought directly from the Pacific coast. Its premium location demanded impeccable presentation, as well as a personality that represented the restaurant's culinary offer: a casual and contemporary reinterpretation of the best food the Mexican coast has to offer.
Identity — The designers developed a casual and relaxed visual language that reminds us of a small Mexican beachfront restaurant, and contrasted it with certain contemporary elements, vintage decorative pieces, and a name that explains the concept in a fresh and concise manner.
Actions — Branding, naming, interior design, copy, advertising campaign for print and digital media.

Design agency: Savvy Studio Client: Costa Nueva

NUEVO LEON, MEXICO

COSTA NUEVA

Mexico's Progressive and Modern Artistic Boom in 1950s

新海岸餐厅 / 墨西哥，新莱昂

墨西哥 20 世纪 50 年代激进而又现代的艺术浪潮

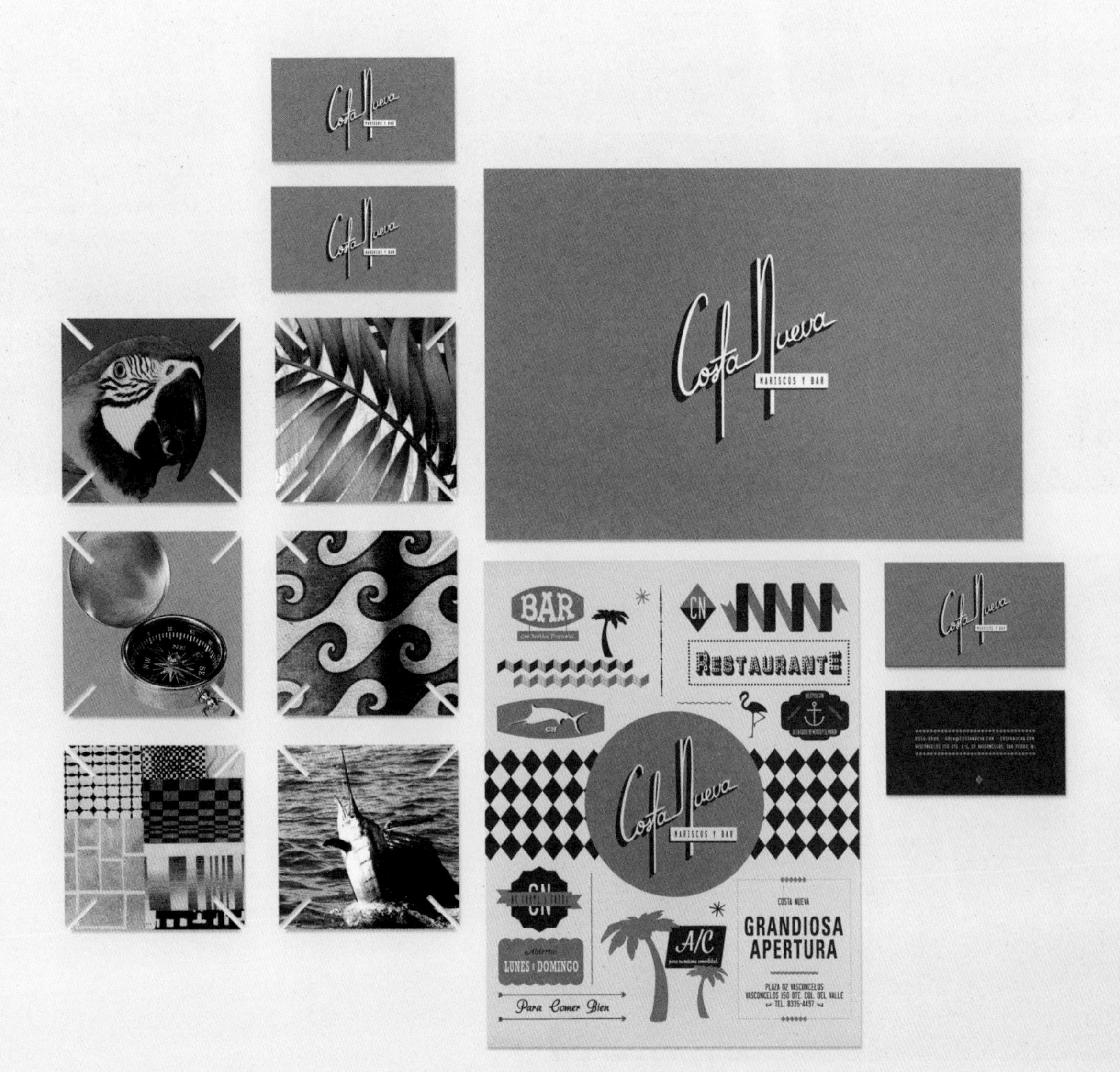

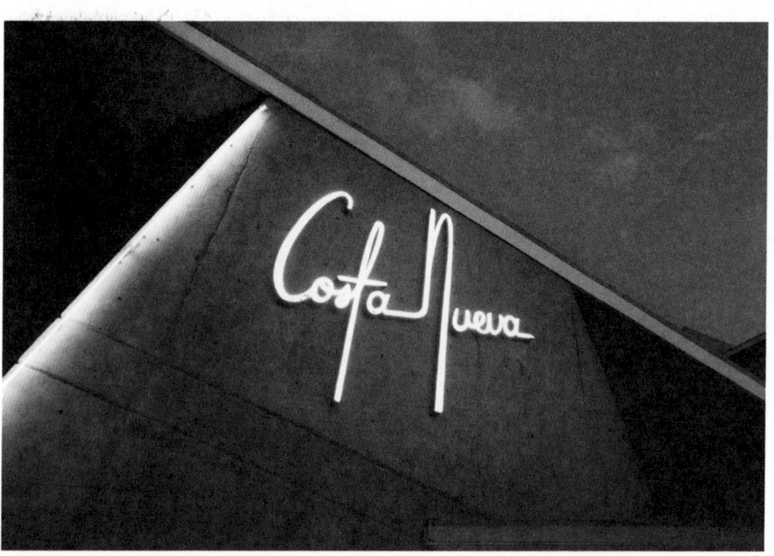

设计分析——新海岸餐厅是一家由两位友人合伙打造的海鲜餐厅，餐厅旨在为食客们提供来自太平洋沿岸最新鲜的食材。优越的地理位置要求匹配一个完美的外观和能够呈现餐厅个性的形象。餐厅的美食以休闲和现代的风格重新诠释了墨西哥海岸的美食。
品牌形象——设计师为餐厅开发了一套轻松的视觉语言，给人以墨西哥海滨餐厅的感觉。他们还添加了一些现代元素、复古装饰与其形成对比。餐厅的名字“新海岸”简洁干脆地显示了新鲜食材的概念。
设计活动——品牌设计、命名、室内设计、复印设计、印刷和数字媒体的广告活动。

设计机构：Savvy 工作室 委托方：新海岸餐饮公司

DE COSTA A COSTA

La Vida es Deliciosa

NUEVO LEON, MEXICO

COSTA CHICA

Most Intuitive Visual Language

小海岸餐厅 / 墨西哥，新莱昂

最直观的视觉语言

BIENVENIDO USTED A:

COSTA CHICA

Breakfast Mariscos

400 Alfonso Reyes L-10
San Pedro — MTY

nalysis — The Costa Chica concept is closely related to that of its older sister Costa Nueva, a fresh seafood restaurant om the Mexican pacific coast. The restaurant itself is a more practical and simple version of its older sibling.
onceptualization — The visual language for Costa Chica derives from the same guiding concept, which emphasizes e freshness of ingredients and gets you as close as possible to the sea without actually being there.
entity — By applying graphic elements in smaller amounts, each one of them acquires a greater importance within e visual composition, a direct analogy of the simplicity, freshness and practicality of Costa Chica.
ction — Branding, naming, print and web advertising.

esign agency: Savvy Studio Client: Costa Nueva

计分析——“小海岸”餐厅的概念贴近它的姐妹餐厅“新海岸”，同样是来自墨西哥太平洋沿岸的海餐厅。这家新餐厅比“新海岸”更加经济实惠，是它的简化版。
念设计——小海岸餐厅的视觉语言源自同样的引导概念，突出了食材的新鲜和贴近海洋的感觉。
牌形象——少量图形元素的应用让每个元素在视觉构成中都获得了更大的影响力，直接呈现了小海岸厅的简洁、新鲜和经济实惠。
计活动——品牌设计、命名、印刷设计和网络推广。

计机构：Savvy 工作室 委托方：新海岸餐饮公司

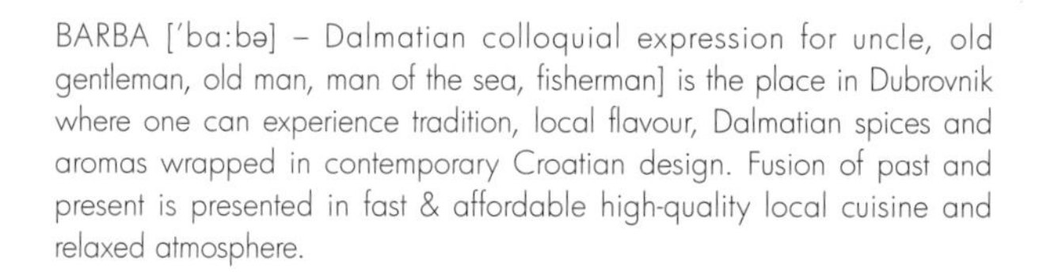

BARBA ['ba:bə] – Dalmatian colloquial expression for uncle, old gentleman, old man, man of the sea, fisherman] is the place in Dubrovnik where one can experience tradition, local flavour, Dalmatian spices and aromas wrapped in contemporary Croatian design. Fusion of past and present is presented in fast & affordable high-quality local cuisine and relaxed atmosphere.

Designers: Negra Nigoević, Filip Pomykalo (graphic design), Marita Bonačić (interior design) Photography: Marita Bonačić Client: BARBA Restaurant

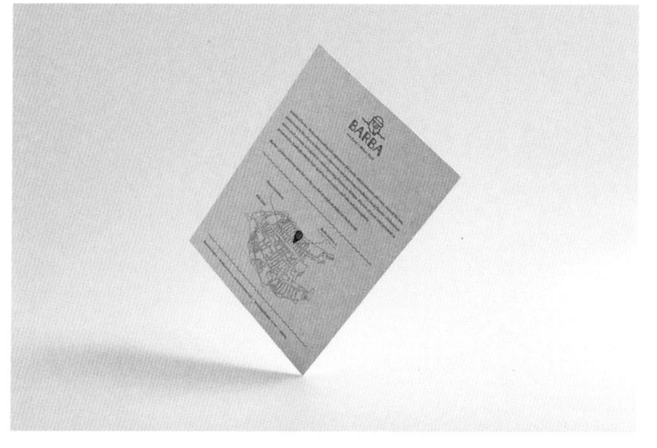

巴尔巴(BARBA)在达尔马西亚口语中用来表达叔叔、老先生、老人、渔夫等意思。巴尔巴餐厅位于杜布罗夫尼克旧城，在那里，你可以在现代克罗地亚设计环境中体验传统和地方风味、达尔马西亚特制调料和香气四溢的美食。过去与现代的融合都呈现在快速、经济的高品质地方美食和轻松的氛围中。

设计师：内格拉·尼格维克、菲利普·波米加罗（平面设计）、玛丽塔·博纳西克（室内设计）摄影：玛丽塔·博纳西克 委托方：巴尔巴餐厅

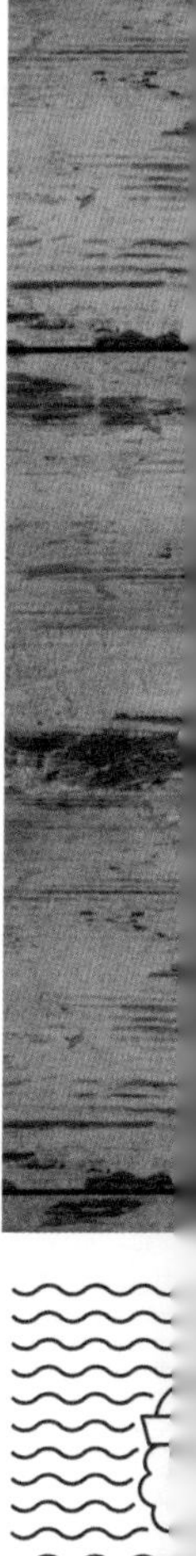

DUBROVNIK, CROATIA

BARBA

A Relaxed Atmosphere of Modern Croatian Design

巴尔巴餐厅 / 克罗地亚，杜布罗夫尼克

现代克罗地亚设计风格营造的轻松氛围

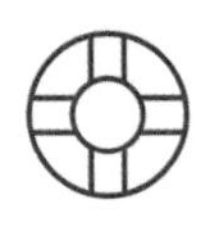

This Brand Identity is for Oxlot 9, an upscale, Southern, seafood restaurant in Covingto Louisiana. The restaurant is located in a unique area of town called the "oxlots," which a set on a diamond pattern. The Oxlot 9 logo is a fish and a nine. The scales of the fish mim the diamond shaped city blocks of the oxlots. The hand-drawn "etching" style speaks to th classic nature of the restaurant and to the history of the area. This logo is inspired by th beauty of the gulf coast, and the history of Covington, LA.

Design agency: Ideogram Designer: Brian Authement Photography: Ideogram Client: OxLot

项目是为奥克斯洛特 9 号餐厅所提供的品牌形象设计。这是一家位于美国路易安那州卡温顿的南方高端海鲜餐厅，餐厅所在的地块名为"奥克斯洛特"，呈菱形餐厅的 LOGO 是一条鱼和一个数字 9，其中鱼鳞模仿了菱形的地块形状。手绘的刻版画风格体现了餐厅的古典内涵和该地区的历史。LOGO 的设计灵感来自于海滨海地区的魅力和卡温顿的历史。

设计机构：**Ideogram 设计公司** 设计师：**布莱恩·奥斯门特** 摄影：**Ideogram 设计公** 委托方：**奥克斯洛特 9 号餐厅**

COVINGTON, USA

OXLOT 9

Etching Style Expresses Classic Nature

奥克斯洛特 9 号餐厅 / 克罗地亚，杜布罗夫尼克

蚀刻版画风格展现古典内涵

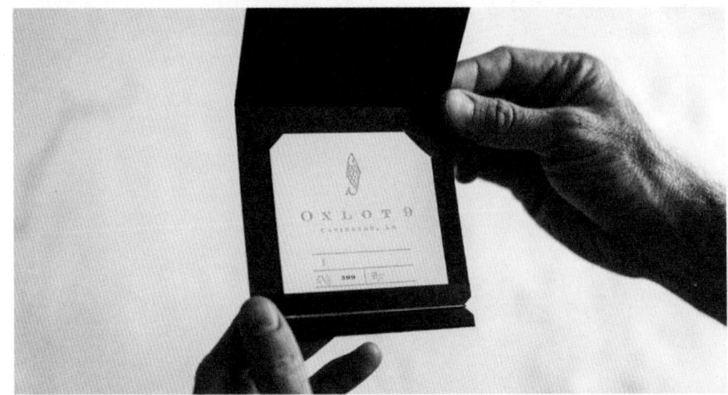

Mains

BISTRO
小酒馆

HONG KONG, CHINA

LA VACHE!

Inspired by French Cartoonist Jean-Jacques Sempé!

好好餐厅 / 中国，香港

灵感来自于法国漫画家让 - 雅克 · 桑贝！

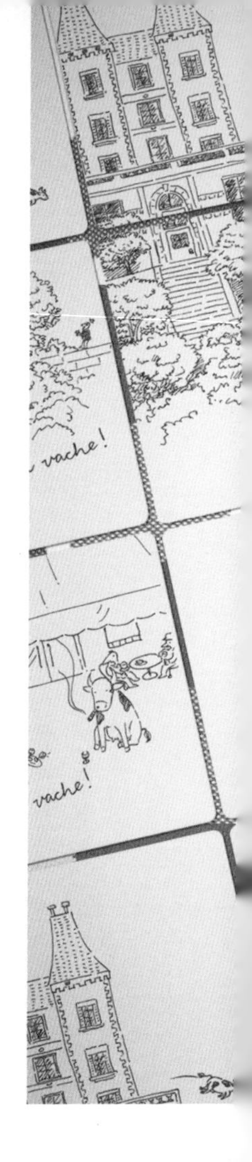

A Parisian bistro in Hong Kong paying homag to classic Parisian steak frites. Inspired by Fren cartoonist Jean-Jacques Sempé, humour is broug to this brasserie's dining experience through t comic illustrations crafted for the identity, depicting series of witty moments featured on menus, coaste matchboxes – even the toilet paper.

Design agency: Substance Executive creati director: Maxime Dautresme Designers: Olivia Che Min Shim, Eleanor Downie Client: Black She Restaurants

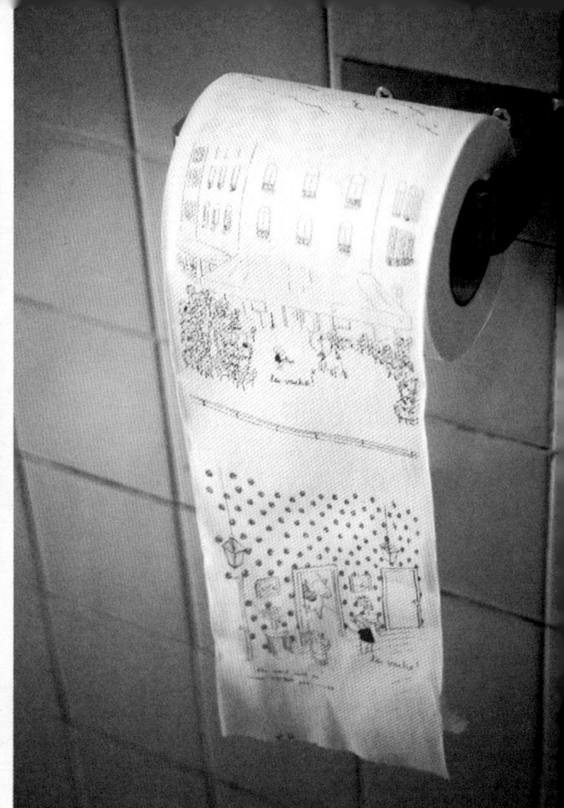

这家位于香港的巴黎酒馆以经典的巴黎牛排薯条为特色菜品。餐厅从法国漫画家让－雅克·桑贝那里获得了灵感，通过品牌形象漫画让幽默遍及整个就餐体验。菜单、杯垫、火柴盒，甚至厕纸上都绘有幽默诙谐的插画。

设计机构：Substance 设计公司 创意总监：马克西姆·多特里斯米 设计师：奥利维亚·陈、沈敏、埃莉诺·唐尼 委托方：黑羊餐饮集团

SINGAPORE

THE RUSTIC BISTRO

Black and White in Deconstructivism and Post-modernism

乡村酒馆 / 新加坡

解构主义与后现代风格里的黑白画映

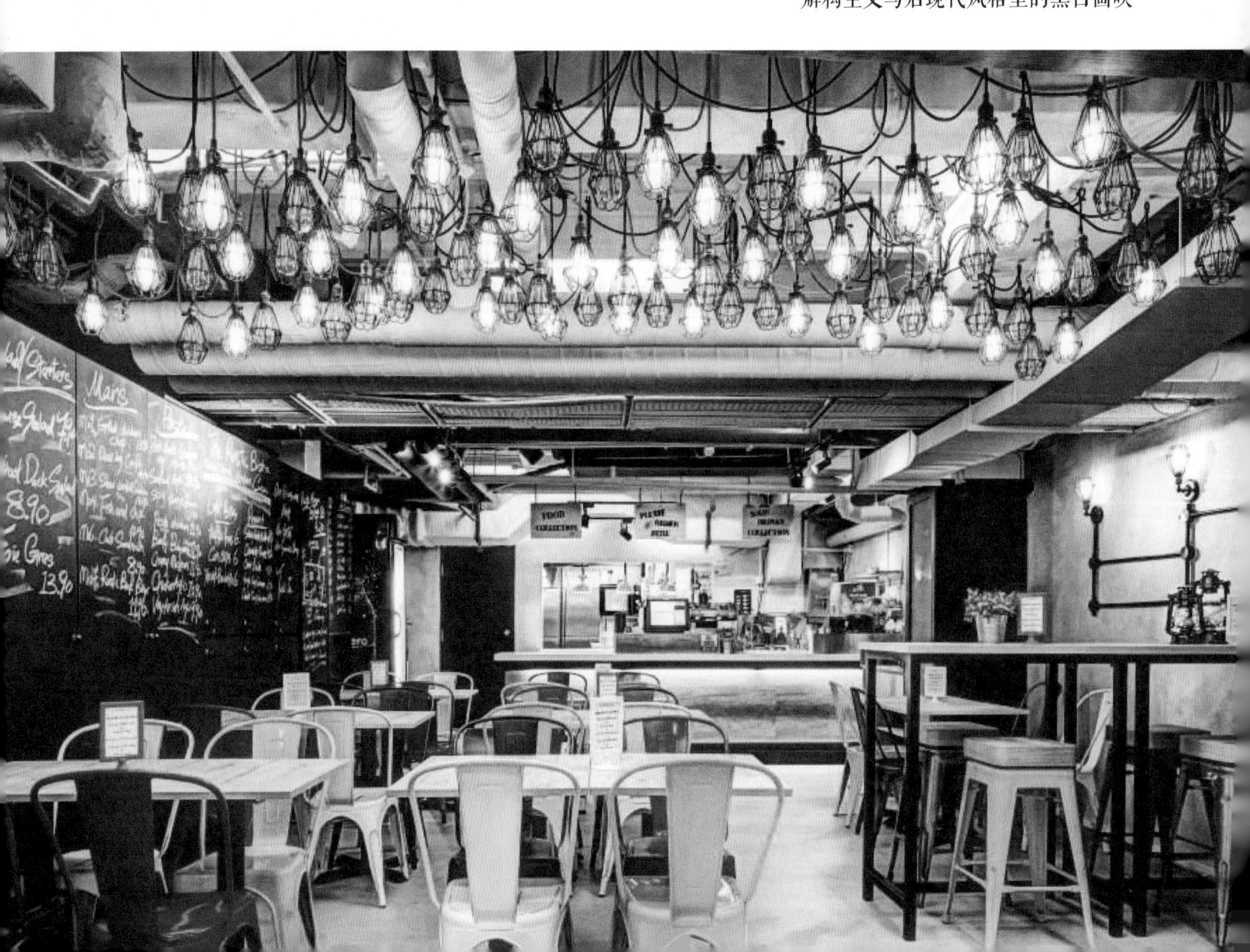

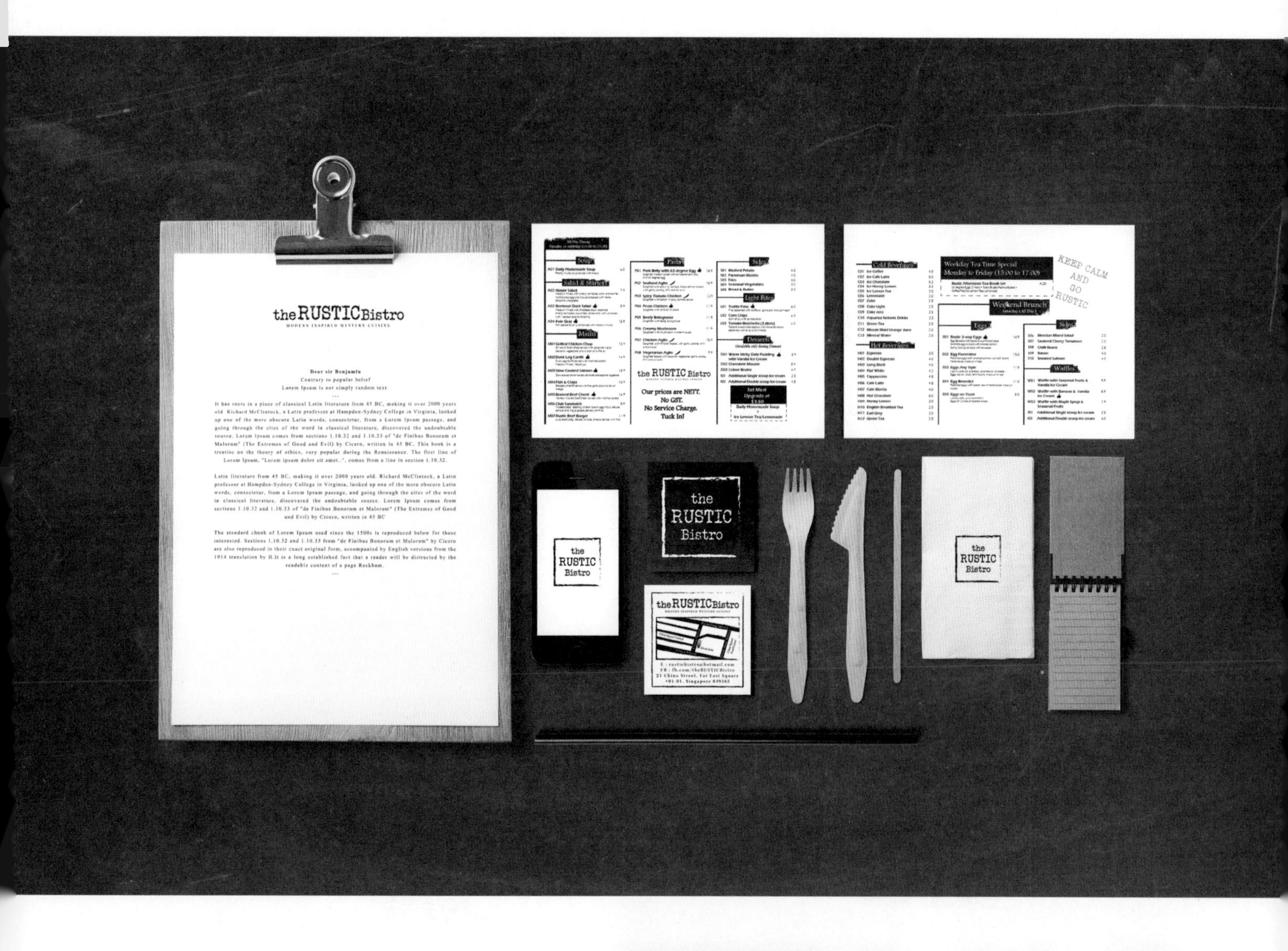

This project is Lemongraphic's first bistro branding project which started in 2014 with the elaboration of identity The RUSTIC Bistro Modern Inspired Western Cuisine. Born out of passion for Modern Inspired Western Cuisine, The Rustic Bistro was the brainchild of Chef Eddy Wan and Chef Stephen Yong. Both chefs helmed more than 10 years of experience in culinary experience in various hotels and chain restaurants.

Design agency: Lemongraphic Designer: Rayz Ong Photography: The Rustic Bistro Client: The Rustic Bistro

本项目是 Lemongraphic 设计公司于 2014 年所打造的第一个酒馆品牌设计项目。乡村酒馆由主厨艾迪·万和主厨斯蒂芬·尹共同经营，他们对现代西方美食有着无比的热情，拥有十多年的烹饪经验，曾就职于各大酒店和连锁餐厅。

设计机构：Lemongraphic 设计公司 设计师：雷斯 · 翁 摄影：乡村酒馆 委托方：乡村酒馆

LAS VEGAS, USA

ENOTECA SAN MARCO

An Eclectic Mix of Typefaces

圣马尔科红酒餐厅 / 美国，拉斯维加斯

字体混合的奇妙设计

The brand design for this project was inspired by the rich cultural heritage of Venice itself — part Eastern, part Liberty, part Art Nouveau. Utilising an eclectic mix of typefaces, it is meant to evoke, not copy, the fantastic design you see everywhere in Venice, particularly in the opulent bars which line the actual Piazza San Marco.

Design agency: Memo Productions, New York City Creative director: Douglas Riccardi Photography: Memo Productions Client: Batali and Bastianich Hospitality Group

项目的品牌设计深受威尼斯丰富的文化历史所影响，有一些东方韵味，一些图书馆的感觉，还有一些新艺术风格。各种字体的混合能令人想起威尼斯随处可见的奇妙设计，特别是街头那些非凡的酒吧。

设计机构：Memo Productions 设计公司 创意总监：道格拉斯・里卡尔迪 摄影：Memo Productions 设计公司 委托方：Batali and Bastianich 酒店服务集团

Libiamo ne' Lieti Calici (aka BRINDISI) From the Opera "La Traviata"

(Giuseppe Verdi | Francesco Maria Piave)

Libiamo

NE' LIETI CALICI, CHE LA BELLEZZA INFIORA:

E la fuggevol ora s'inebriì a voluttà.

LIBIAM NE' DOLCI FREMITI CHE SUSCITA L'AMORE,

Poichè quell'occhio

al core onnipotente va.

Libiamo

AMORE, AMOR FRA I CALIÇI PIÙ CALDI BACI AVRÀ.

Let's drink from our merry glasses adorned with beauty; and to this fleeting hour intoxicated with pleasure. Let's drink with the sweet tremblings of love, for love's omnipotent eye pierces down to the bottom of the heart.

Let's drink to love,

for love's kisses are hotter with wine in our glasses.

DIO NON HA CREATO CHE L'ACQUA.

L'uomo ha fatto il Vino.

God only created the water. It was man who made Wine.

Cin Cin!

Alla Salute!

(2 Italian toasts)

bottiglia: 1. Sciampagnotta | 2. Borgognona | 3. Renana | 4. Bordolese | 5. Collo | 6. Fondo

The Venetian® LAS VEGAS
3355 Las Vegas Blvd. South
enotecasanmarco.com
702·677·3390

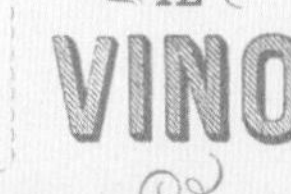

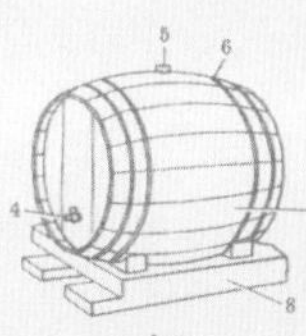

1. Vite | 2. Madrevite | 3. Gabbia | 4. Canella | 5. Zaffo | 6. Cerchio | 7. Doga | 8. Sedile

IT WILL BE MACCHERONI

· I SWEAR TO YOU ·

that will

UNITE ITALY! *

The Venetian® LAS VEGAS
3355 Las Vegas Blvd. South
enotecasanmarco.com
702·677·3390

1 RAVIOLI
Cremona claims to have invented them, though some would argue it was Genoa.

2 AGNOLOTTI
Piedmont-style ravioli dating back to 1798, today most commonly filled with meat.

3 ALL'AMATRICIANA
The name refers to the town of Amatrice, and contains guanciale, onion, chili peppers, tomato and cheese.

4 FETTUCCINE ALL'ALFREDO
More commonly called *fettuccine al burro*, it is made with butter and parmigiano and was made famous at Rome's Alfredo alla Scrofa in 1914.

5 ALLA CARBONARA
"Charcoal style." Roman-style preparation of spaghetti tossed with pancetta, parmigiano and raw egg.

6 MALLOREDUS (or *Small Balls*)
Tiny gnocchi-like dumplings from Sardinia, made with semolina flour, saffron and salt.

7 PENNE (or *Little Quills*)
A specialty of Campania, they take their name from the Latin word for feather, *penna*.

Pesto alla Genovese
3 tbsp Pine Nuts
2 cups Basil Leaves
1 clove Garlic
½ cup Olive Oil
¼ cup Parmigiano
Combine, and make into a paste.

Aglio e Olio
4 tbsp. Olive Oil
3 cloves Garlic
1 Peperoncino
1 tbsp. Fresh Parsley
Heat oil, add garlic and peperoncino, do not let it burn. Toss with cooked pasta and parsley. Salt and pepper to taste.

* Giuseppe Garibaldi, on liberating Naples in 1860

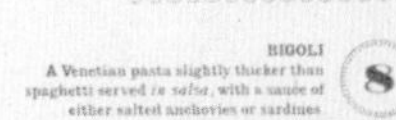

8 BIGOLI
A Venetian pasta slightly thicker than spaghetti served *in salsa*, with a sauce of either salted anchovies or sardines.

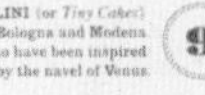

9 TORTELLINI (or *Tiny Cakes*)
A specialty of Bologna and Modena. The shape is said to have been inspired by the navel of Venus.

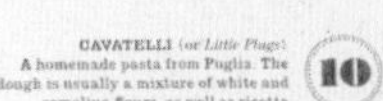

10 CAVATELLI (or *Little Plugs*)
A homemade pasta from Puglia. The dough is usually a mixture of white and semolina flours, as well as ricotta.

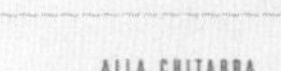

11 ALLA CHITARRA
"Guitar style." A fresh egg pasta from Abruzzo, cut on a stringed instrument called a *chitarra* that results in a four-sided noodle.

12 ALLA PUTTANESCA
"In the Style of Harlots." A spicy Calabrese tomato sauce believed to have originated in brothels because it was quick and easy to make between appointments.

13 ALLA NORMA
A Sicilian sauce made of tomato, eggplant and *ricotta salata*. It is believed to have been named after the opera "Norma" by Vincenzo Bellini, though some say it merely means "pasta made in the normal way."

GLOSSARIO: Cappellini: *little hairs* | Farfalle: *butterflies* | Fettuccine: *little ribbons* | Gemelli: *twins* | Linguine: *little tongues* | Orecchiette: *little ears* | Spaghetti: *little strings* | Stelline: *little stars* | Vermicelli: *little worms* | Ziti: *bridegrooms*

SOME SAY THE WORD MACARONI COMES FROM THE SICILIAN "MACCARONI" MEANING MADE INTO A DOUGH BY FORCE.

Gnocchi

THERE ARE ABOUT 150 VARIETIES OF PASTA, AND MORE THAN 600 PASTA SHAPES PRODUCED WORLDWIDE.

Sedani

PROPERLY COOKED PASTA SHOULD BE "AL DENTE" (LITERALLY TO THE TOOTH), A BIT FIRM, BUT TENDER.

Conchiglie

ITALIANS SAY THAT YOU CAN JUDGE A PERSON'S CHARACTER BY THE WAY THEY EAT SPAGHETTI.

Agnolotti

UNTIL THE INVENTION OF TOMATO SAUCE IN THE MID 18TH CENTURY, ITALIANS ATE PASTA WITH THEIR HANDS.

Lumache

ACCORDING TO MISS MANNERS, A FORK IS THE ONLY IMPLEMENT TO BE USED WHEN EATING SPAGHETTI.

habibis

ISLAMIC RESTAURANT

清真餐厅

SAN PEDRO GARZA GARCIA, MEXICO

HABIBIS

An Arabic-Mexican Fusion Taqueria Using Gentle Typeface to Show Friendliness

哈比比斯餐厅 / 墨西哥，圣佩德罗加尔萨加西亚

柔和字体打造亲切感的清真餐厅

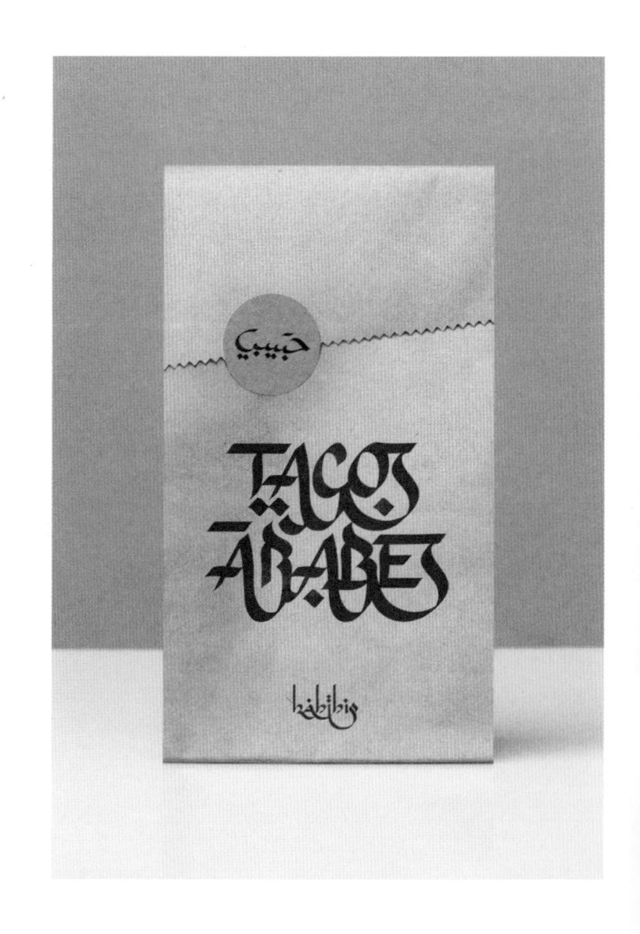

Habibis is an Arabic-Mexican fusion taqueria located in San Pedro Garza García, a city enriched by the culinary treats of its third generation Arab immigrants. Previously a humble taco stand, Habibis approached the designers with the task of creating a brand that communicated the foods' exceptional mixed background and quality without losing its street-friendly and casual demeanor. The proposal is a brand that adapts stylised Arabic calligraphy to a typical Mexican street setting. Complete with neon colours and inexpensive materials, like craft paper bags. Deep research and careful understanding of the Arabic alphabet was needed to design, using calligraphic pens and special brushes, the various words and signage in both Arabic and Latin. The custom type is accompanied by Gotham, a gentle and neutral typeface that would allow the bespoken logotypes to stand out above everything else. The pattern is based on traditional keffiyeh (a Middle Eastern headdress fashioned from a square scarf) and gorgeously intrinsic mosaic patterns.

Design agency: Anagrama Client: Habibis

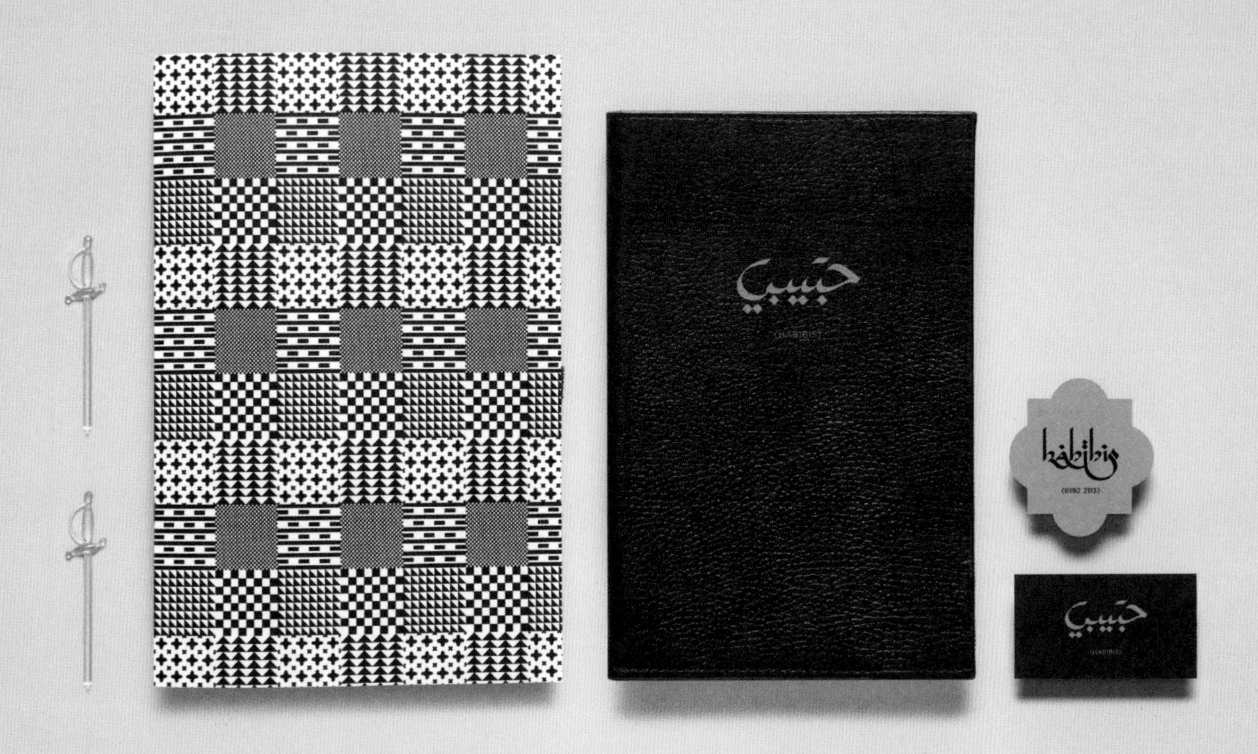
حبيبي
(HABIBIS)
habibis

TACOS ARABES
habibis
ENTRADAS
TACOS
ESPECIALIDADES
EXTRAS
POSTRES
BEBIDAS
habibis
JOCOQUE

GRACIAS

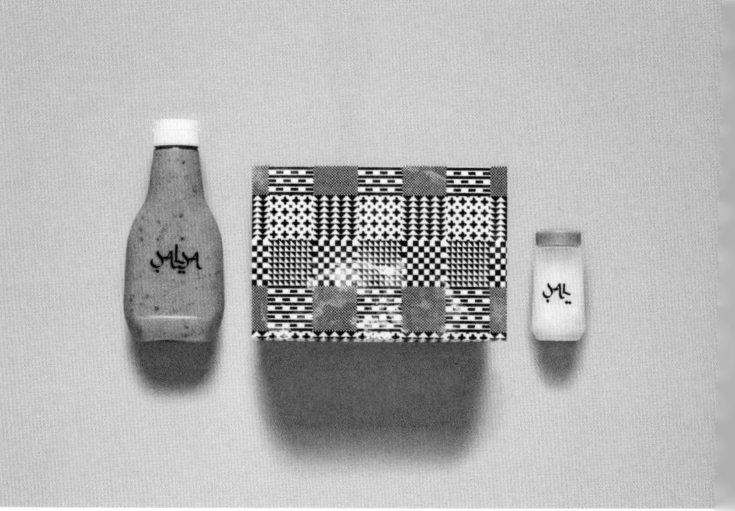

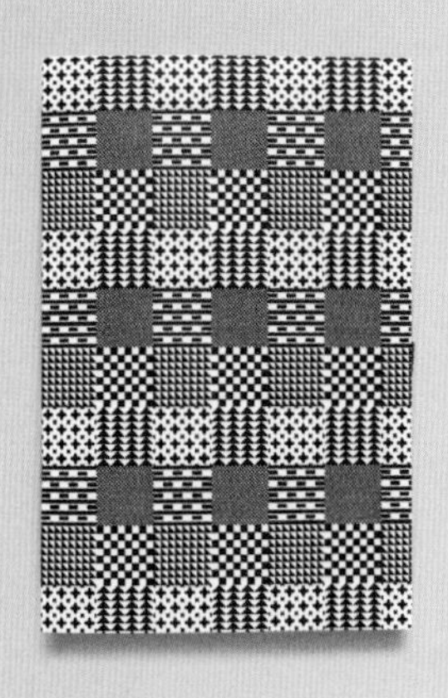

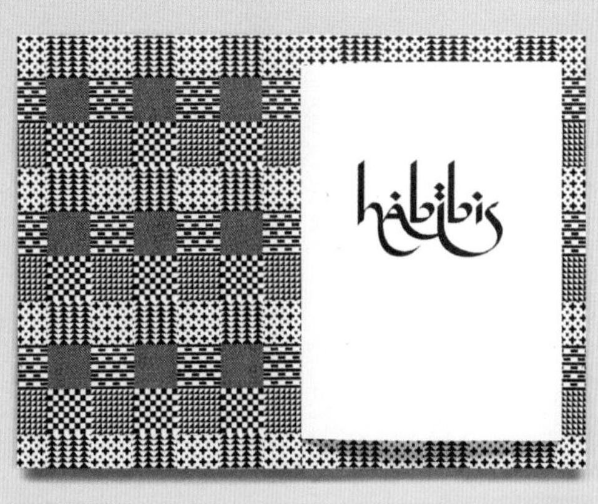

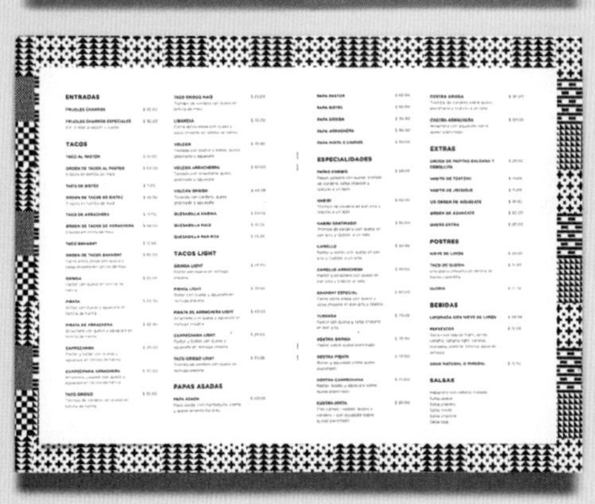

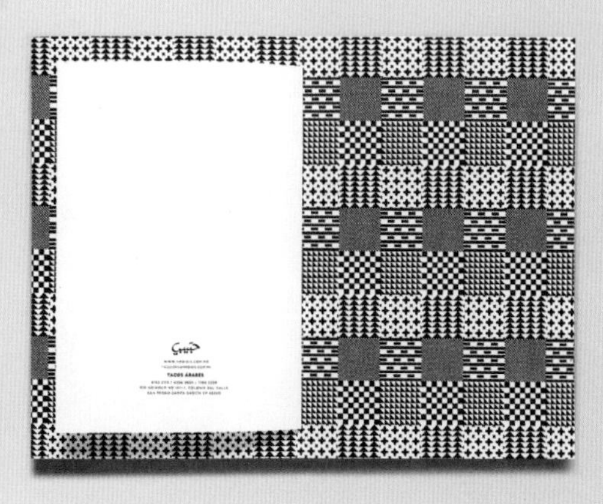

哈比比斯是一家阿拉伯/墨西哥融合餐厅，位于圣佩德罗加尔萨加西亚——一座拥有大量阿拉伯移民的城市。哈比比斯餐厅之前只是一个小摊，贩卖墨西哥煎玉米卷。现在他们希望设计师为其打造一个品牌，既要体现美食独特的融合背景和品质，又不能失去从前那种街头小吃的亲切感。最终的品牌设计将阿拉伯书法与典型的墨西哥街头场景融为一体，辅以荧光色与牛皮纸等物美价廉的材料。设计深入研究并了解了阿拉伯字母，用书法笔和特殊的刷子以阿拉伯文和拉丁文书写了各种单词和标识。柔和而中性的 Gotham 字体令 LOGO 标志脱颖而出。餐厅的图案设计从传统的阿拉伯头巾以及华丽的镶嵌图案中获得了灵感。

设计机构：Anagrama 设计公司 委托方：哈比比斯餐厅

Design and art direction of various collaterals for Serai. Serai is a family-owned restaurant that offers Asian and Western flavours. The interior design is a work of Kerusi.

Design agency: Koyoox Photography: Koyoox Client: Serai

项目为客栈餐厅提供了全方位的设计和艺术指导。客栈餐厅是一家私家餐厅，专营东西方美食。餐厅的室内设计由 Kerusi 提供。

设计机构：**Koyoox 设计公司** 摄影：**Koyoox 设计公司** 委托方：**客栈餐厅**

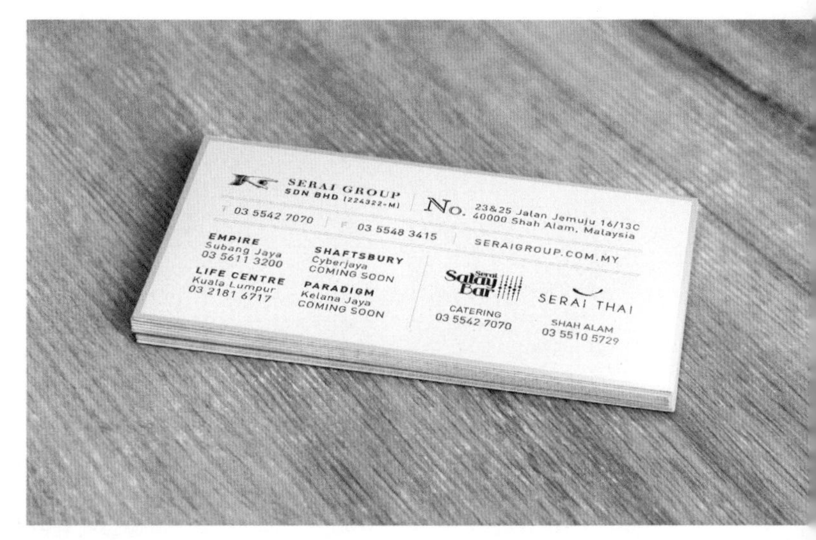

KUALA LUMPUR, MALAYSIA

SERAI

"Simple yet Incomparably Beautiful"

客栈餐厅 / 马来西亚，吉隆坡

"朴素而天下莫能与之争美"

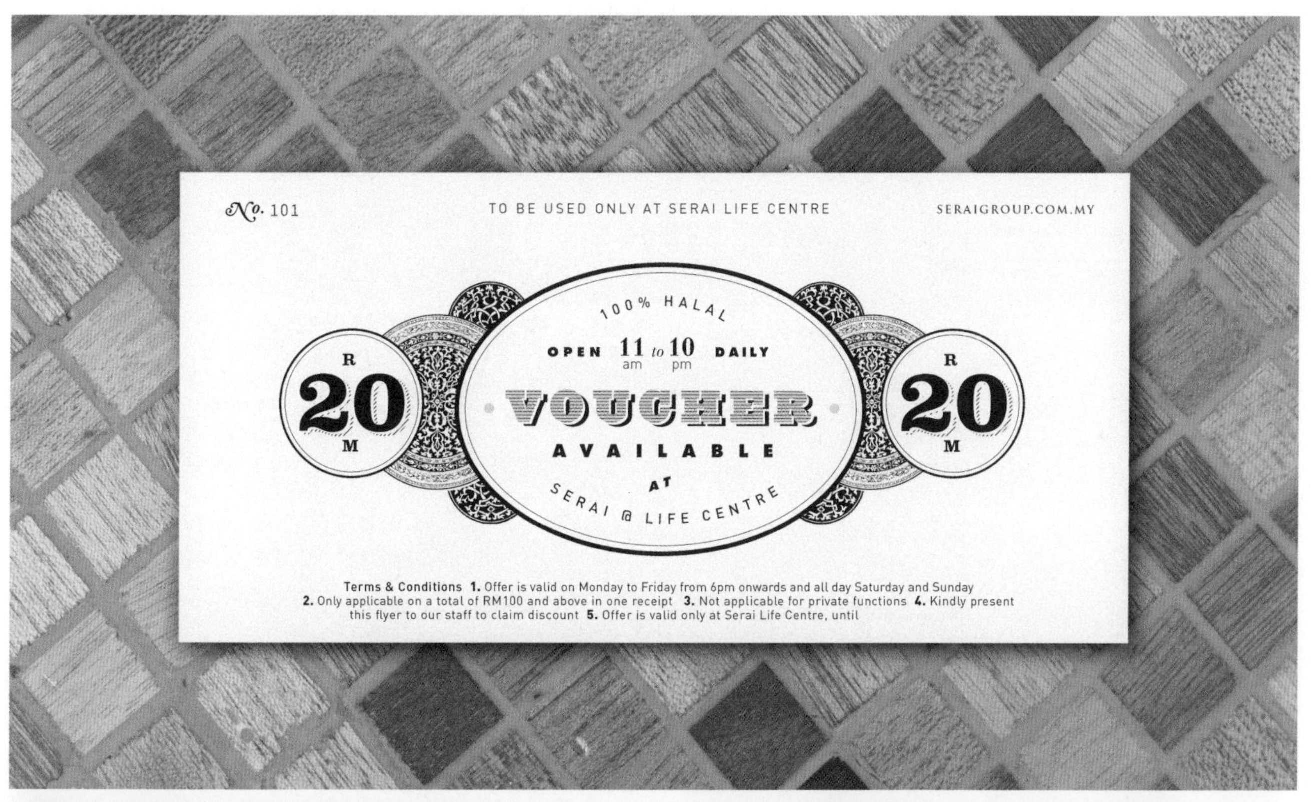

Cuando
todo
haga
BOOM
quiero
estar
contigo
DESDE 2005
BAOBAB
COCINA VEGETARIANA

VEGETARIAN
RESTAURANT
素食餐厅

MADRID, SPAIN

RESTAURANTE BAOBAB

A Vegetarian Restaurant with Humanism

猴面包餐厅 / 西班牙，马德里

人本主义的素食餐厅

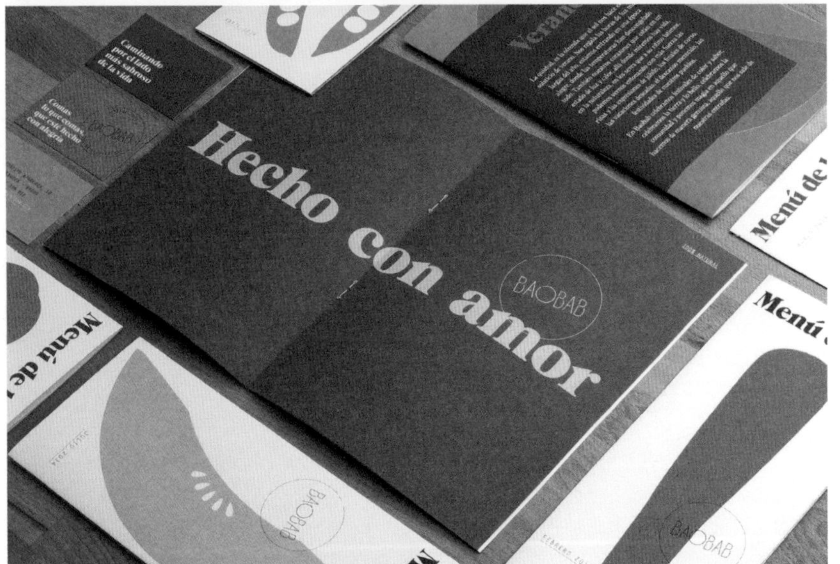

Baobab is a vegetarian restaurant providing healthy food and 100% natural without dogmas or complex. Baobab's new look is the result of a way of understanding life and food, which seeks the complicity of the client, spiced with humor and humanism. An invitation to enjoy the food, colour, reading, music and good company.

Design agency: Tropical Designers: Álex Sánchez & Sergio Palao Photography: Álex Sánchez & Sergio Palao Client: Restaurante Baobab

猴面包餐厅是一家素食餐厅，提供健康食品和100%天然无添加美食。猴面包餐厅新形象的设计灵感来自于对生命和食物的理解，设计从消费者处寻求共鸣，到处都显露出幽默感和人本主义。设计吸引着人们享受美食、色彩、阅读、音乐和好伴侣。

设计机构：Tropical 设计公司 设计师：艾利克斯·桑切斯、塞尔吉奥·帕劳 摄影：艾利克斯·桑切斯、塞尔吉奥·帕劳 委托方：猴面包餐厅

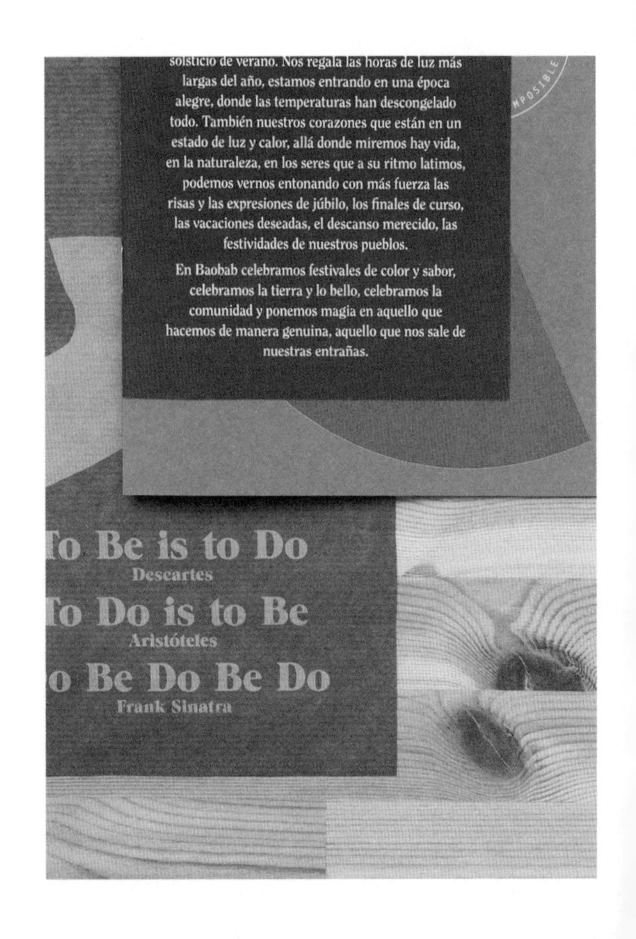

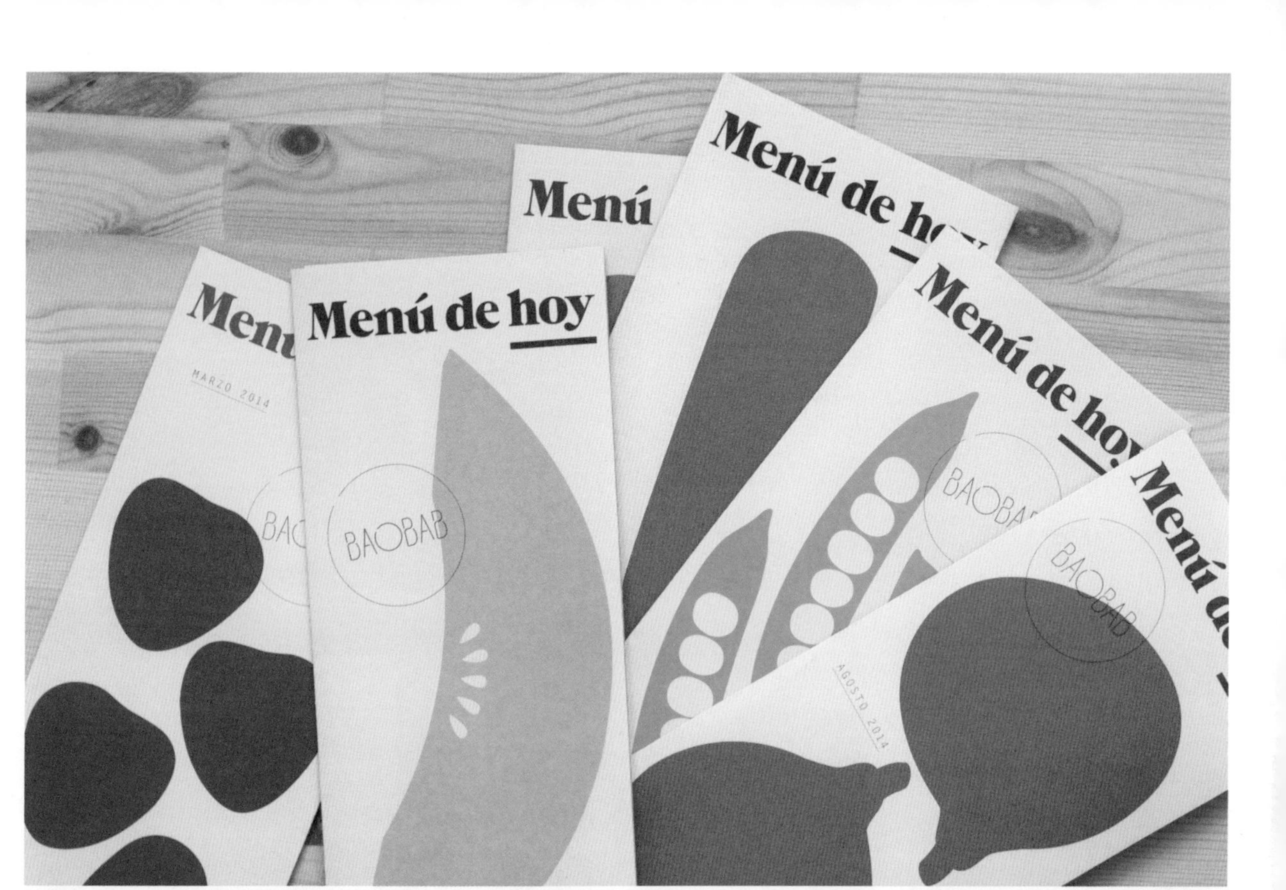
Menú de hoy
MARZO 2014
BAOBAB
AGOSTO 2014

Menú de hoy
BAOBAB
ABRIL 2014

Verano
La quietud, es la ofrenda que el sol nos hace en el solsticio de verano. Nos regala las horas de luz más largas del año, estamos entrando en una época alegre, donde las temperaturas han descongelado todo. También nuestros corazones que están en un estado de luz y calor, allá donde miremos hay vida, en la naturaleza, en los seres que a su ritmo latimos, podemos vernos entonando con más fuerza las risas y las expresiones de júbilo, los finales de curso, las vacaciones deseadas, el descanso merecido, las festividades de nuestros pueblos.
En Baobab celebramos festivales de color y sabor, celebramos la tierra y lo bello, celebramos la comunidad y ponemos magia en aquello que hacemos de manera genuina, aquello que nos sale de nuestras entrañas.

100% NATURAL
SANO
SABROSO
VEGETARIANO
BAOBAB

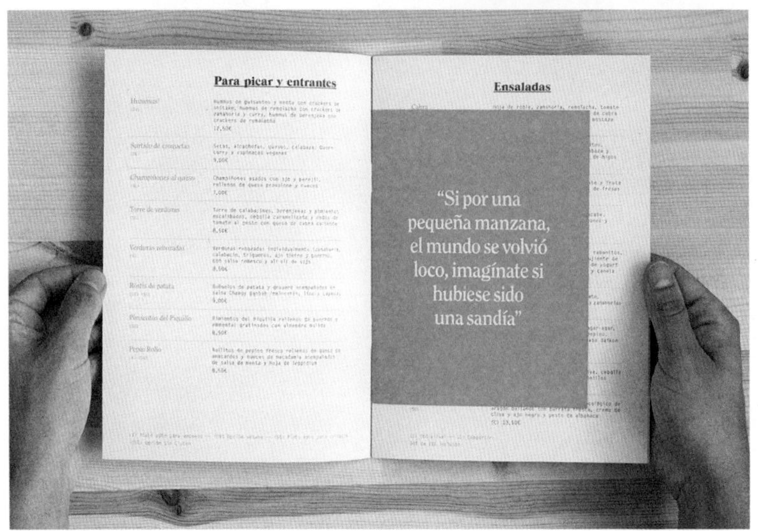
Para picar y entrantes
Ensaladas
"Si por una pequeña manzana, el mundo se volvió loco, imagínate si hubiese sido una sandía"

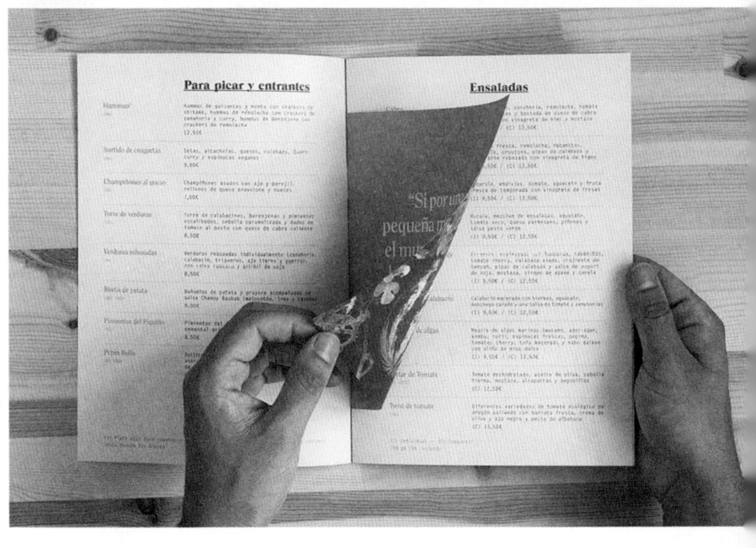
Para picar y entrantes
Ensaladas

MENÚ DE HOY
976 359 011
CARTA DE VERANO
BAOBAB
SUBIR

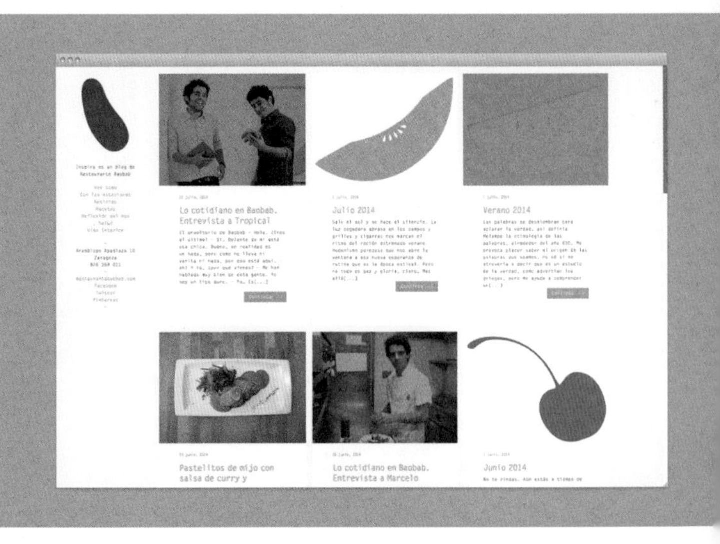
Lo cotidiano en Baobab. Entrevista a Tropical
Julio 2014
Verano 2014
Pastelitos de mijo con salsa de curry y
Lo cotidiano en Baobab. Entrevista a Marcelo
Junio 2014

Arzobispo Apaolaza 10
Zaragoza 50009
T 976 359 011
HORARIO
Lunes a Domingo
de 13:00 a 16:00h
y Martes a Sábado
de 21:00 a 23:30h
RESTAURANTEBAOBAB.COM
Busca InspiraBaobab en las redes sociales y:
Comparte
Escucha
Charla
Mira
BAOBAB
100% NATURALES
DESDE 2005

Me gustaría ser todo de vino y beberme yo mismo

FAST-FOOD RESTAURANT

快餐店

KIEV, UKRAINE

SIMPLE

All Details Sustain and Complement Each Other

简单餐厅 / 乌克兰，基辅

极致细节下的相辅相成

The task was complex: in addition to the name, logo and corporate identity Brandon had to develop an interior design. The designers focused on a close-knit team working with the architect Anna Domovesova. They started at the same time, analysing, checking and adjusting to each other's steps and global course in general. This approach gave Brandon an opportunity to get an integral product, where all details sustain and complement each other.

Design agency: Brandon Creative directors: Boris Alexandrov, Anna Domovesova Designers: Olga Novikova, Anton Storozhev Illustrator: Elena Parhisenko Client: Anton Gusakov

simple.
жду
гостей
7
Простота
со Вкусом
be Simple
eat Simple

#1
#2
#3

ПРИСАЖИВАЙТЕСЬ...

项目的设计任务比较复杂，除了命名、LOGO 设计和企业形象设计之外，
andon 设计公司还必须与建筑师安娜· 多姆维索瓦共同合作，开发一套室
设计方案。同时，设计师还必须分析、检查并调整每一步设计，实现整体
标。设计的过程让 Brandon 有机会打造了完整的产品链，使得所有细节都
够相辅相成。

计机构：Brandon 设计公司 创意总监：波利斯 · 亚历山德罗夫、安娜 · 多
索维瓦 设计师：奥尔加·诺维科娃、安东·斯托洛泽夫 插画设计：艾琳娜·帕
西森科 委托方：安东 · 古萨科夫

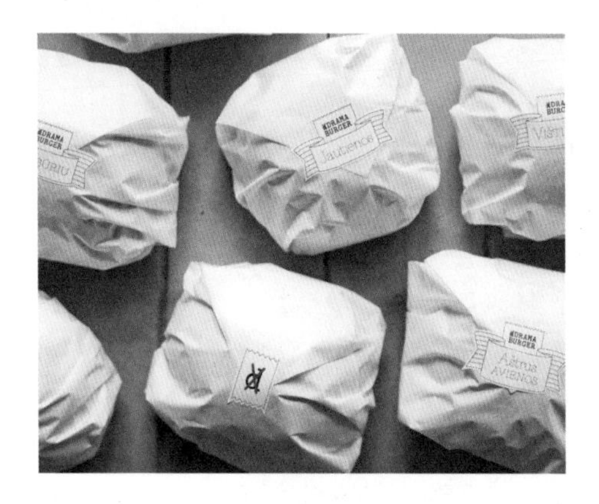

OVERLAND PARK, USA

DRAMA BURGER

A Dramatic Burger Restaurant

戏剧汉堡 / 美国，欧弗兰帕克

戏剧化效果的独特汉堡店

ere are fast food burgers, there are burgers and
n there's new trend of gourmet burgers, where
at is differently grilled, bun is home cooked,
redients are crazier and more tasty. A few well
velled men decided to open such place in Vilnius.
t for the city. They needed some fresh name and
h approach to how burger joints look.

sign agency: New Agency Designer: Lina
rcinonyte Illustrator: David Schiesser Photography:
rius Petrulaitis Client: UAB Septyni mesainiai

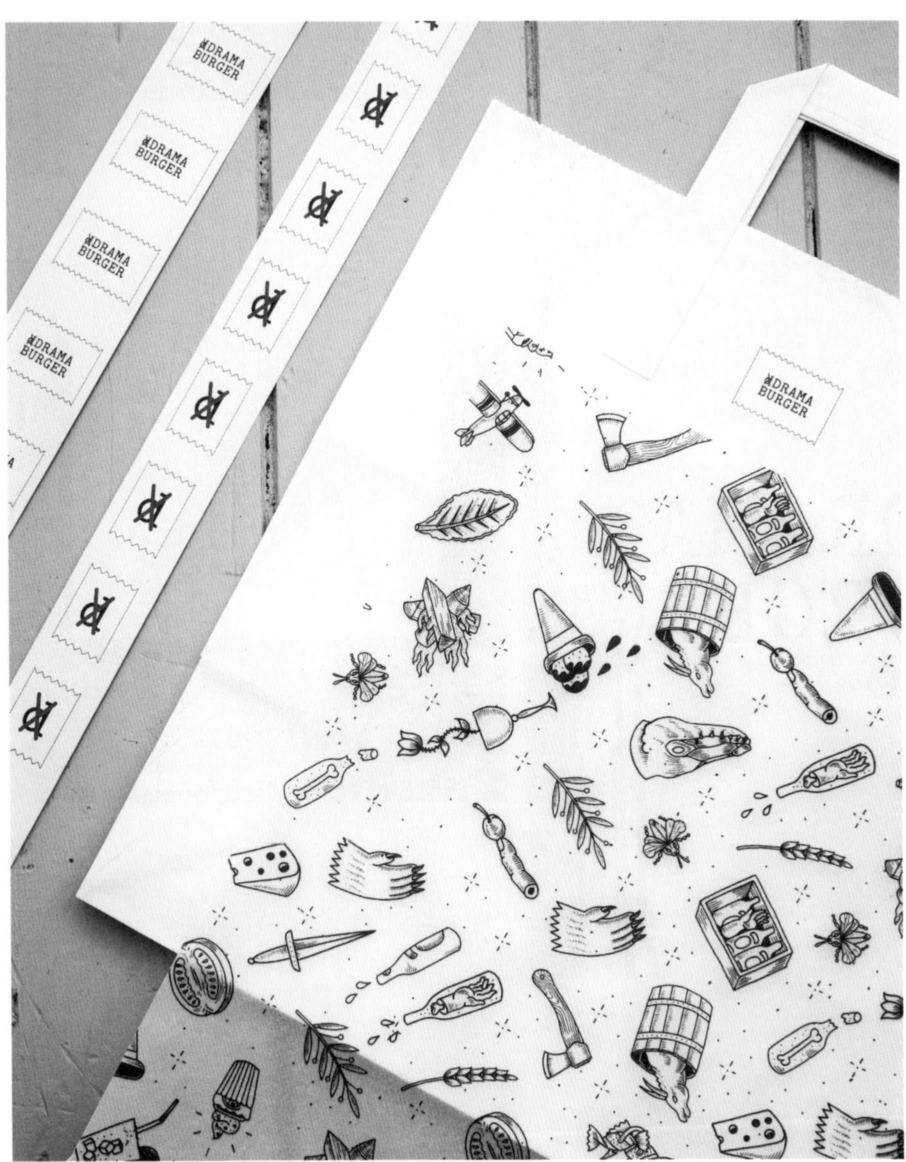

Enter DRAMA BURGER, the name we created. "Drama", because there's so much going in the burger - the drama of taste. And "Drama", because you'll sweat yourself picking up one from huge selection. We designed the logo, so it reflects the drama - letter crossed out, all nervous. We wrote slogans, saying that it's damn hard to choose between eating now or later, lamb or chorizo, hands or knives & forks. And we've decided to draw helluva illustrations.

汉堡又可分为快餐汉堡、普通汉堡和新潮美味汉堡，后者的肉饼采用独特的煎制方法，面包是自制的，配料丰富，更加美味。几位曾游历四方的人士决定在维尔纽斯地区开一家独特的汉堡店。他们需要新鲜的名字和新鲜的设计来搭配美味的汉堡。设计师将汉堡店命名为“戏剧汉堡”。取名“戏剧”是因为汉堡的口味具有戏剧效果，十分丰富。顾客很难从琳琅满目的菜单中选出一款汉堡。品牌 LOGO 的设计同样反映了戏剧化效果——被划掉的字母给人以紧张感。设计师撰写了宣传口号：“现在吃还是等会儿吃？羊肉还是香肠？手抓还是刀叉？太难选了！”并且决定以各种插画作为装饰。

设计机构：New Agency 设计公司 设计师：丽娜 · 玛尔希诺伊特 插画设计：大卫 · 席塞尔
摄影：达利斯 · 佩德鲁莱提斯 委托方：UAB Septyni mesainiai 公司

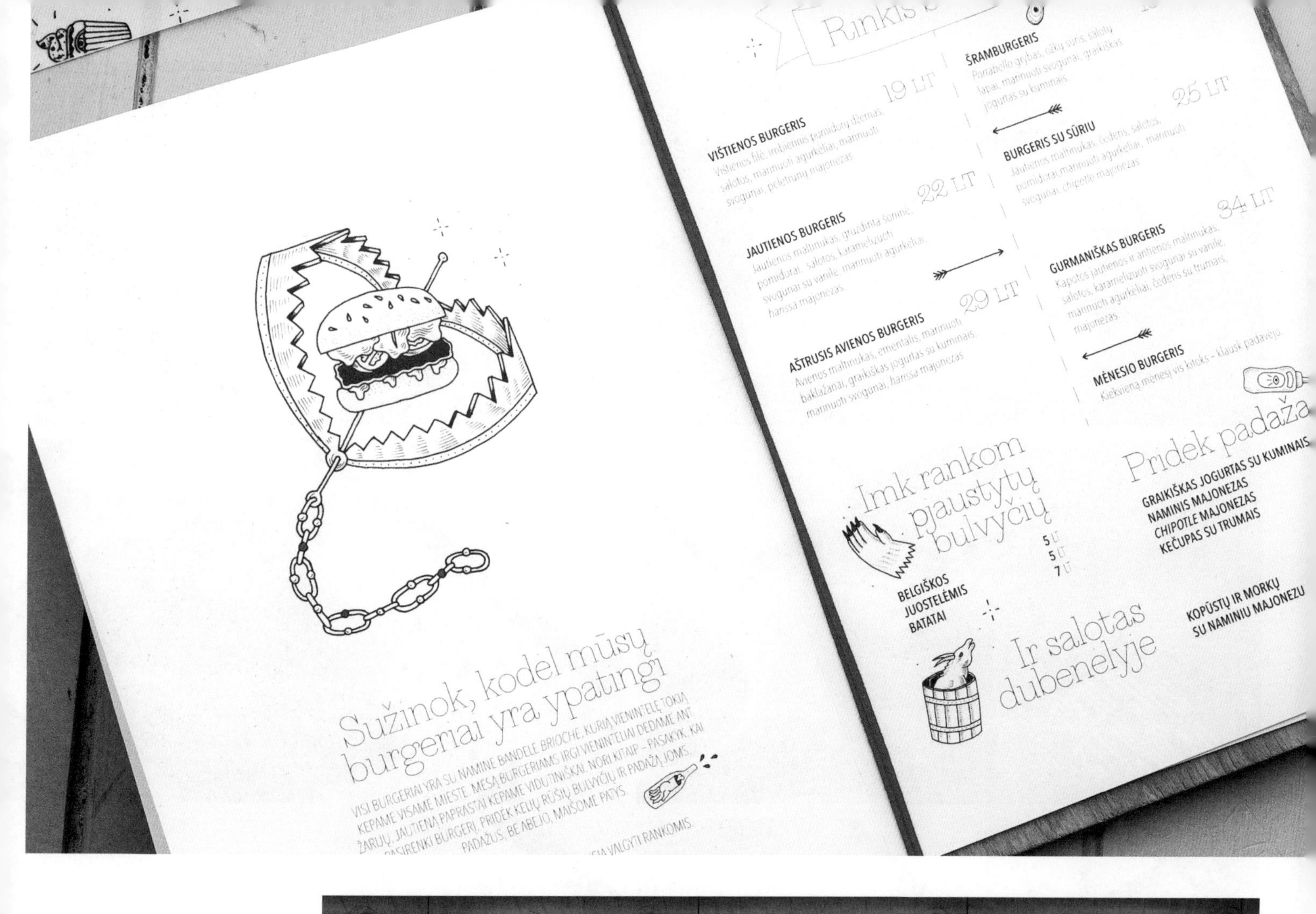
Sužinok, kodel mūsų burgeriai yra ypatingi
ŠRAMBURGERIS
VIŠTIENOS BURGERIS
19 LT
BURGERIS SU SŪRIU
25 LT
JAUTIENOS BURGERIS
22 LT
GURMANIŠKAS BURGERIS
34 LT
AŠTRUSIS AVIENOS BURGERIS
29 LT
MĖNESIO BURGERIS
Imk rankom pjaustytų bulvyčių
BELGIŠKOS
JUOSTELĖMIS
BATATAI
5 LT
5 LT
7 LT
Pridek padažą
GRAIKIŠKAS JOGURTAS SU KUMINAIS
NAMINIS MAJONEZAS
CHIPOTLE MAJONEZAS
KEČUPAS SU TRUMAIS
Ir salotas dubenelyje
KOPŪSTŲ IR MORKŲ SU NAMINIU MAJONEZU

DRAMA

Branding and Art Direction for a chicken wing food truck based in New York.

Design agency: Ageless Galaxy Photography: Ageless Galaxy Client: Wing'n It

项目是为纽约的一家专门供应鸡翅美食的快餐车提供的品牌形象设计和艺术指导。

设计机构：Ageless Galaxy 设计公司 摄影：Ageless Galaxy 设计公司 委托方：飞翔餐饮公司

NEW YORK, USA

WING'N IT NYC

Freestyle Typeface

纽约飞翔餐车 / 美国，纽约

自由“字”在

ATLANTA, USA

CHICK-A-BIDDY

Bright Illustrations for Fast Food

小鸡餐厅 / 美国，亚特兰大

明亮的快餐插画

Chick-a-Biddy Farm Fresh Chicken & Sides is a southern cooking inspired chicken restaurant with a twist. Located in Atlanta, GA, the overall concept rooted in southern cuisine, is never pretentious, always delicious. Chick-a-Biddy strives to serve food that is raised properly and is in fact better for us. Chick-a-Biddy just as the name suggests is well crafted southern cooking which is exactly what the owner wanted their brand to reflect. Brand components consisted of; naming, logo system, brand identity, menus, interior and exterior signage, environmental graphics and apparel.

Design agency: Carpenter Collective Designer: Tad Carpenter
Photography: Carpenter Collective

Chick-a-Biddy
FARM FRESH CHICKEN & SIDES
Desserts
EXIT

Biddy
DON'T BE A
CHICKEN!
Eat the

鸡农场新鲜鸡肉餐厅是一家别具风味的南方美食餐厅。餐厅位于美国乔治亚州的亚特兰大，整概念扎根于美国南方美食，从不做作，只求美味。小鸡餐厅力求为人们提供最好的美食，餐厅名字很好地反映了它的南方烹饪特色。餐厅的品牌设计包括命名、LOGO 系统、品牌形象、菜单、内外标识、环境图形和服装设计。

计机构：Carpenter Collective 设计公司 设计师：泰德 · 卡彭特 摄影：Carpenter Collective 设计公司

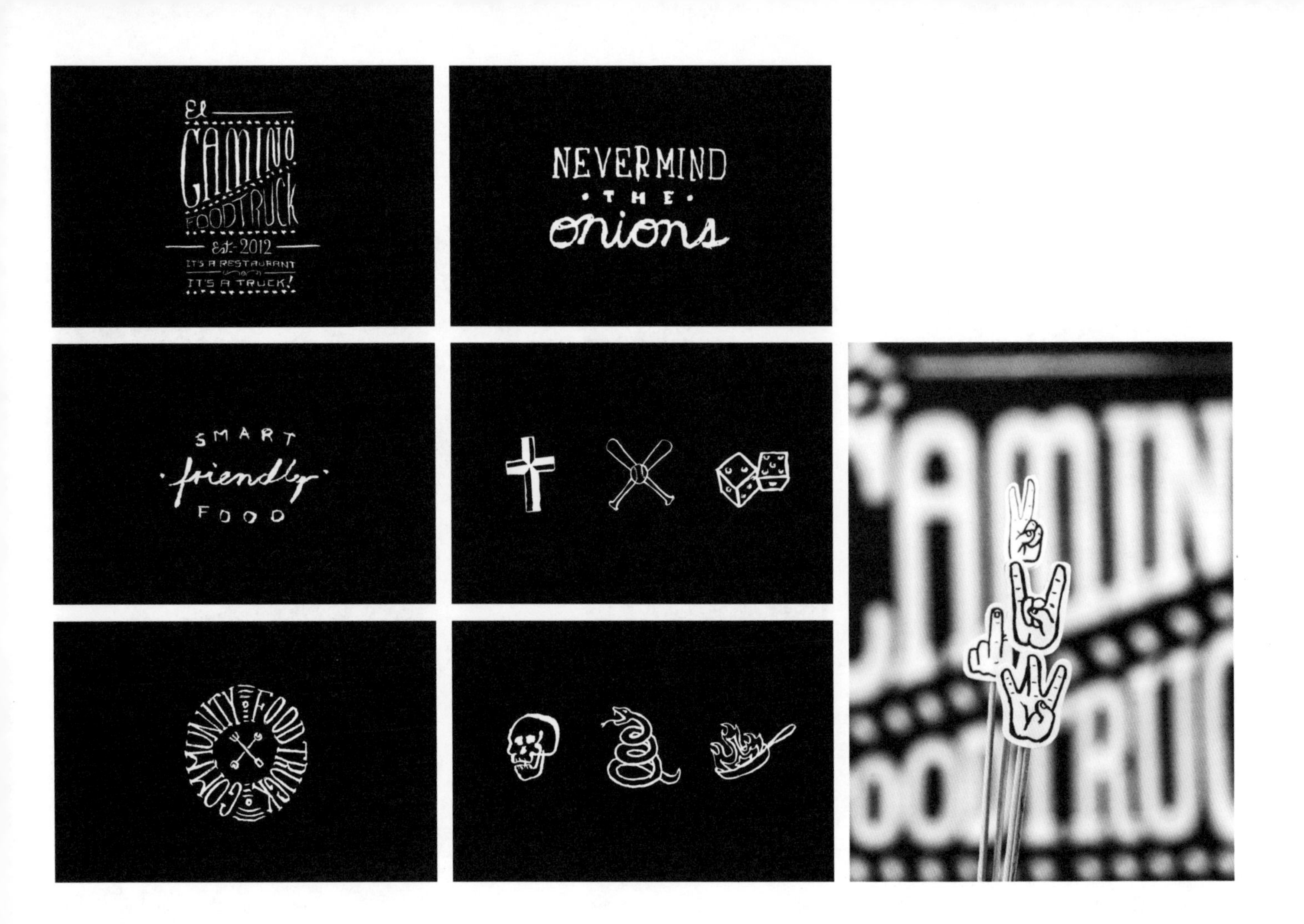

SAN PEDRO GARZA GARCIA, MEXICO

EL CAMINO

Sustainable Graphic Language

埃尔·卡米诺餐车 / 墨西哥，圣佩德罗加尔萨加西亚

可持续图形语言

ɔr this project the designers developed an exhaustive graphic language which constantly growing. The aim is to make the Foodtruck very recognisable rough its easily identifiable graphic style. Americana and biker tattoos are vo strong influences, expressing the truck's rough and Texan personality in an ppealing way. El Camino Foodtruck is a tribute to an artisanal approach, ɔm the meticulous preparation of the food it serves to the creation and plementation of its visual language that was conceived in pencil and paper nd later hand drawn directly on the truck.

esign agency: Savvy Studio Photography: Alejandro Cartagena Client: El amino

计师为餐厅开发了一套独特且能持续进化的图形语言，其目标是通高辨识度的图形风格使餐车能脱颖而出。美式乡村刺青和骑行者刺是两个最重要的元素，体现了餐车的强硬感和得克萨斯风情。埃尔·卡诺餐车决定向手工工艺致敬：厨师一丝不苟地制作食物；设计师用笔绘制图形语言，然后将其直接手绘在餐车上。

计机构：Savvy 工作室 摄影：亚力山卓·卡塔赫纳 委托方：埃尔·卡诺餐饮公司

LONDON, UK

BYRON

Hand-drawn Type and Illustrations Catering for the Restaurant

拜伦餐厅 / 英国，伦敦

手绘文字和插画迎合餐厅氛围

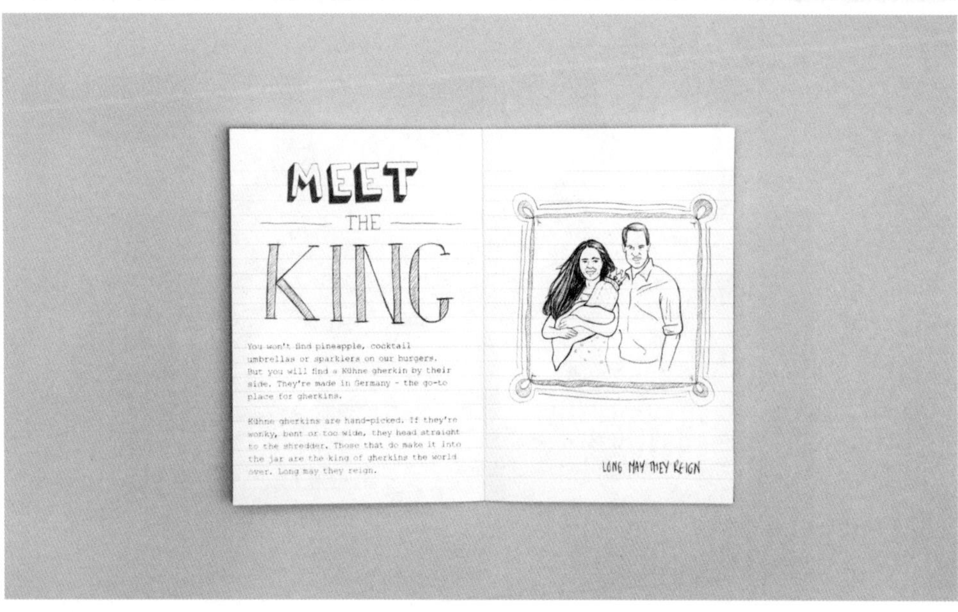

& SMITH worked with We All Need Words to help Byron explain the ins-and-outs of the brand to their staff. Staying away from a typical 'brand book', the designers kept it light-hearted and not too serious. They used hand drawn type and illustrations and created something that wouldn't look out of place in one of their restaurants.

Design agency: & SMITH Designer: Sam Kang Photography: & SMITH

& SMITH 设计公司与 We All Need Words 设计公司帮助拜伦餐厅打造了一套独特的品牌形象规划。不同于传统的“品牌手册”，设计师让这套品牌形象规划保持着轻松愉快的基调。他们用手绘文字和插画打造了一套与餐厅整体氛围相一致的设计。

设计机构：& SMITH 设计公司 设计师：山姆·康 摄影：& SMITH 设计公司

LONDON, UK

SMACK LOBSTER ROLL

Inspired by Signage on Boats

斯迈克龙虾卷餐厅 / 英国，伦敦

灵感来自船体手绘标识

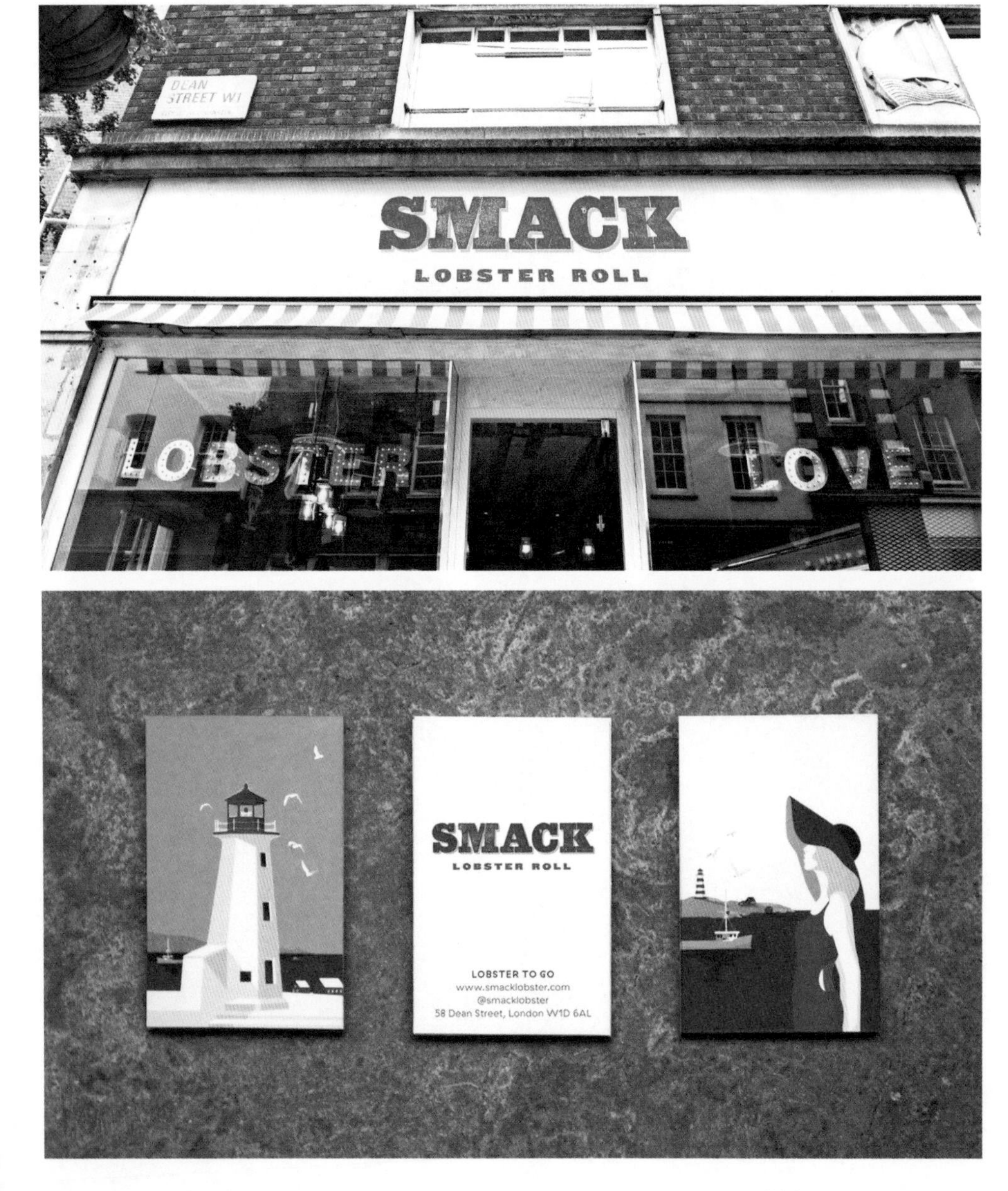

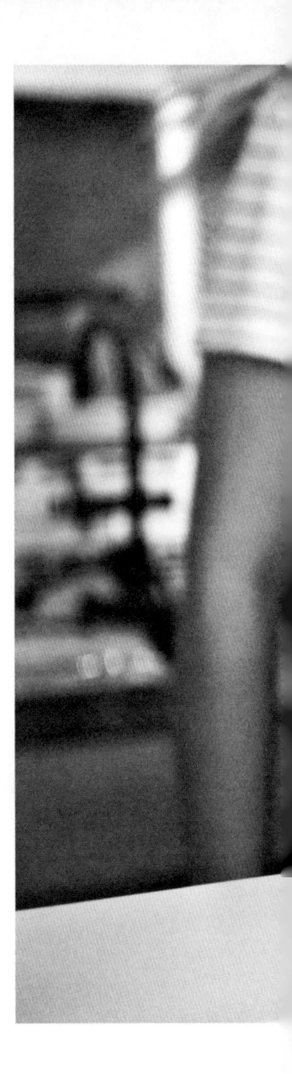

& SMITH were brought on board to rebrand Smack Deli under its new name of Smack Lobster Roll. This included a new logo, positioning, all POS, website and illustration. Inspiration for the logo came from hand painted signage on 'smack' boats which are used for lobster fishing.

Design agency: & SMITH Designer: Sam Kang
Photography: & SMITH

& SMITH 设计公司受委托为斯迈克餐饮集团下的斯迈克龙虾卷餐厅打造品牌形象。整体设计包含新 LOGO、品牌定位、销售终端设计、网站和插画。LOGO 的设计灵感来自于捕龙虾的小船上的手绘标识。

设计机构：& SMITH 设计公司 设计师：山姆·康
摄影：& SMITH 设计公司

KANSAS, USA

UNFORKED

A Playful, Distinct Language to Create a High-quality Fast Food Brand

无刀叉餐厅 / 美国，堪萨斯

独特有趣的设计语言打造高品质快餐品牌

The owners of Sheridan's Frozen Custard set out to UNdo what people believe about fast food. Design Ranch's goal was to communicate that Unforked is a socially responsible, high-quality alternative to a quick meal. Through fresh graphics and a playful, distinct language, the designers did just that. Collaborating with 360 Architecture on the restaurant's interiors, the designers were able to name and brand the restaurant, design menus, to-go bags, employee uniforms and everything in between.

Design agency: Design Ranch Client: Sheridan's Unforked

谢里丹软香乳冻的经营者决心颠覆人们对快餐的想法。Design Ranch 设计公司的目标是将无刀叉餐厅打造成富有社会责任感的高品质快餐品牌。设计师利用清新的图形设计和独特有趣的设计语言实现了这一点。他们与 360 建筑事务所共同合作完成了餐厅的室内设计，并成功地设计了餐厅的名字、品牌形象、菜单、外卖袋、员工制服等。

设计机构：Design Ranch 设计公司 委托方：无刀叉餐厅

BRUSSELS, BELGIUM

LE PARI'S

Lovely Patterns and Fresh Style

帕里快餐厅 / 比利时，布鲁塞尔

可爱图案演绎清新风格

"Le Pari's" is a restaurant / fast food located in Brussels, Belgium. Its concept is to bring the culinary elements of France in a modern, fresh and friendly atmosphere. Total branding.

Design agency: fhancquart.com Designer: Florent Hancquart Photography: Florent Hancquart Client: Le Pari's

帕里快餐厅位于比利时的首都布鲁塞尔。餐厅的理念是在现代、清新、友好的氛围中为食客提供经典的法式美食。项目呈现了餐厅的全套品牌规划设计。

设计机构：www.fhancquart.com 设计师：弗洛伦特·汉克加特 摄影：弗洛伦特·汉克加特 委托方：帕里餐厅

Le Pari's

Holly
BURGER
DONOSTIA ‡ SAN SEBASTIÁN

SAN SEBASTIÁN, SPAIN

HOLLY BURGER

Inspired by the Banana Leaf Wallpaper of the Beverly Hills Hotel in Los Angeles

奥利汉堡 / 西班牙，圣塞巴斯蒂安

灵感来自美国洛杉矶比弗利山酒店的香蕉叶墙纸

ly Burger is the coolest new burger restaurant in San Sebastián, a small yet, beautiful town at the heart he Basque Country, in the north of Spain. The restaurant is named after Holly who was the aunt of the aurant owner, Iñigo Otegui. She is the one responsible for many of the secret burger recipes present in their cious menu. The idea was to create a real American-style brand with a fresh mix of style references. The gners' first inspiration came from various vintage, hand-drawn American typographies present in old shop dows and a banana leaf wallpaper that had been originally designed in 1942 by decorator Don Loper for Beverly Hills hotel in Los Angeles, California.

gners: Rodrigo Aguadé & Manuel Astorga Client: Holly Burger

奥利汉堡是西班牙北部巴斯克地区小镇圣塞巴斯蒂安最具风格的汉堡餐厅。餐厅是以老板姑姑的名字“奥利”命名的，她为餐厅贡献了许多独特的汉堡制作秘方。餐厅的品牌设计理念是打造一个正宗的美式品牌，同时融入一些清新的时尚风格。设计师的灵感来自于老商店橱窗上各种复古的手绘美国字体和装饰设计师唐·洛佩尔为美国洛杉矶比弗利山酒店所设计的香蕉叶墙纸。

设计师：罗德里格·阿加德、曼纽尔·阿斯托加 委托方：奥利汉堡

Pixelarte's job was to design and develop the whole global corporate identity, including branding, packaging and guidance through the design and decoration process for The Fitzgerald Burger Company, a brand new burger restaurant placed in Valencia. A gourmet fast-food restaurant in a different space, where taste delicious burgers and hotdogs flame grill cooked with high quality fresh ingredients. For the branding creation process, the designers did a hand drawn lettering with brushpen, and also tailor-made illustrations for the product icons to build an original image, a little bit badass but with a premium vintage touch. A different place that brings us directly to New York's Soho heart.

Design agency: Pixelarte Creative director: Jorge Timoteo Designer: Josep Navarro Client: The Fitgerald Burger Co.

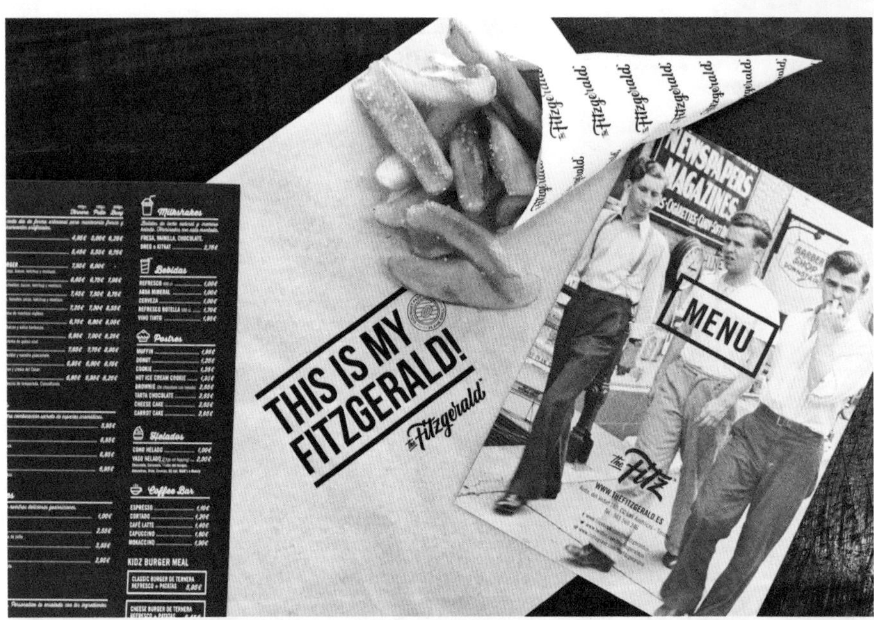

Pixelarte的任务是为巴伦西亚新晋的菲茨杰拉德汉堡餐厅打造全的全球企业形象，包括品牌形象、包装、设计和装饰指导。这家众不同的快餐厅用高品质的新鲜食材为人们提供美味的汉堡和狗。在品牌设计中，设计师用毛笔手绘了文字并量身打造了产品标插画，整体设计有点另类，又有点复古。菲茨杰拉德汉堡餐厅够让我们产生置身于纽约市中心的错觉。

设计机构：Pixelarte设计公司 创意总监：乔治·蒂莫特奥 设计师约瑟夫·纳瓦罗 委托方：菲茨杰拉德汉堡公司

VALENCIA, SPAIN

THE FITZGERALD BURGER CO.

A Premium Vintage Touch by Hand-drawn Lettering with Brushpen

菲茨杰拉德汉堡餐厅 / 西班牙，巴伦西亚

毛笔手绘文字与图标插画所缔造的另类复古风格

LONDON, UK

EMBUTIQUE

Hipstery London Style

艾姆布提克餐厅 / 英国，伦敦

时髦的伦敦风格

Designed for a London restaurant that serves Spanish style sandwiches, Xabier Ogando created Embutique's branding with a clash of heavy typography and bold illustrations, mixed up with a little hipstery London style. The tradition of Spanish products fuses with a London 'hipster' touch. The font and illustrations represent the distinctive identity of the brand.

Designer: Xabier Ogando Client: Embutique

设计师哈比尔·欧甘多为伦敦一家专门提供西班牙风味三明治的餐厅打造了全套的品牌设计。艾姆布提克餐厅的品牌形象由粗大的字体和大胆的插画构成，辅以一些时髦的伦敦风格，为传统的西班牙美食注入了伦敦的时尚感。字体和插画设计体现了该品牌的独特形象。

设计师：哈比尔·欧甘多 委托方：艾姆布提克餐厅

The Bite is a gourmet hamburger restaurant located in Zürich, Switzerland. The iden of the brand emphasises contemporary handmade craftsmanship, along with restaurant's approach to using high-quality ingredients. The tone expresses a fun a friendly attitude inspired by American hamburger culture, in a modest contempor voice. All aspects of the identity were designed bespoke to communicate The Bit values. Stationery, menus, posters, invites, and custom signage made from indust materials compliment the atmosphere of the interior space. All items were printed in traditions of fine printing at OK Haller Druck AG, Zürich.

Design agency: Nychuk Design Designer: Josh Nychuk Photography: Josh Nyc Client: The Bite

一口餐厅是一家位于瑞士首都苏黎世的汉堡餐厅。餐厅的品牌形象突出了代手工工艺和餐厅所使用的高品质食材。设计基调体现了一种开心友好的度，就像美国汉堡文化一样。品牌形象的各个方面都体现了一口餐厅的价值文具、菜单、海报、邀请函和定制的标识都具有浓重的工业气息，与室内间的氛围相互映衬。所有材料全部在苏黎世本地的 OK Haller Druck 公司以传的精美印刷方式印制。

设计机构: Nychuk 设计公司 设计师: 乔什·尼查克 摄影: 乔什·尼查克 委托一口餐厅

ZURICH, SWITZERLAND

THE BITE

Strong Industrial Feeling in Black and White

一口餐厅 / 瑞士，苏黎世

黑白色彩下的浓重工业气息

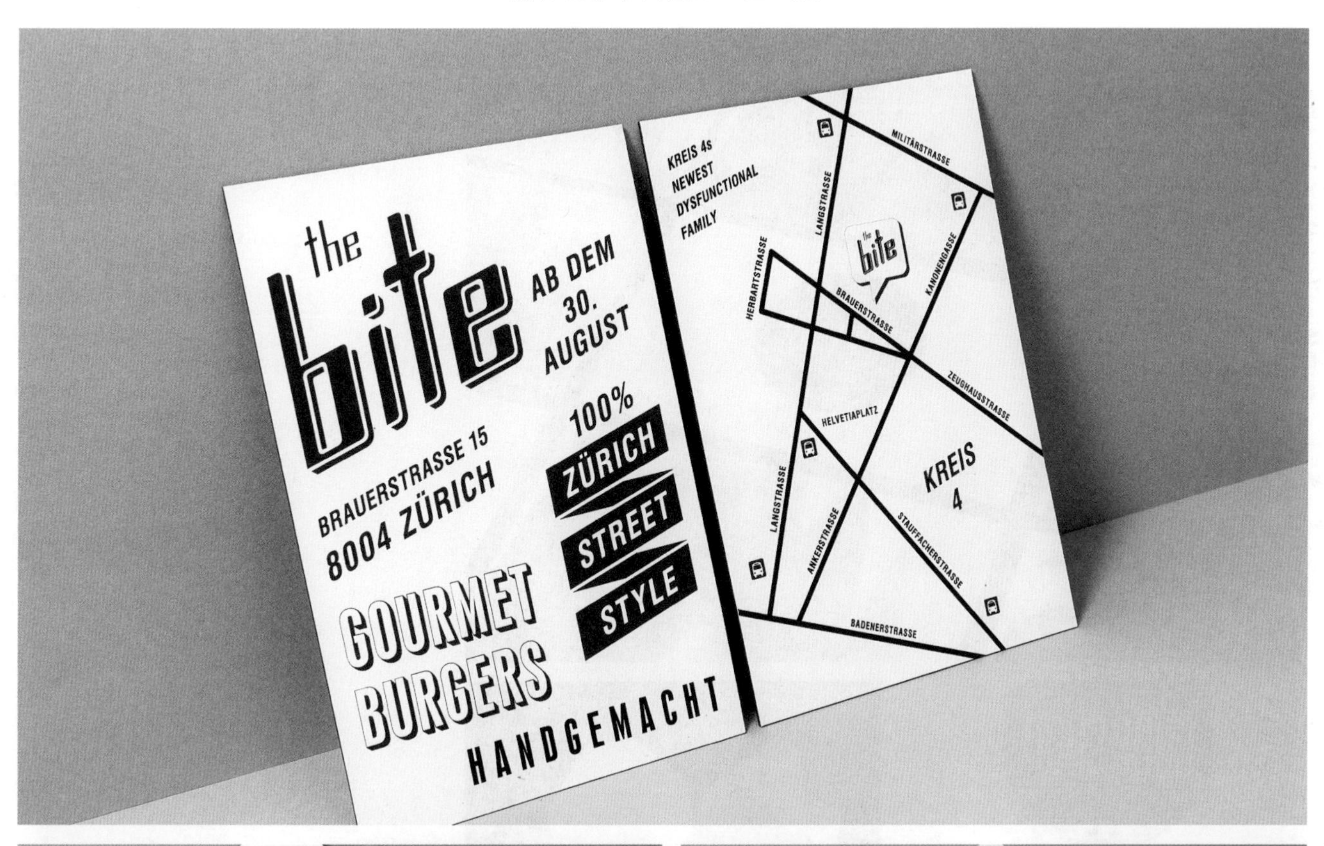

TEL AVIV, ISRAEL

CHOP CHOP

Collage in Small Space

乔普餐厅 / 以色列，特拉维夫

拼贴画融入小空间

Chop Chop is an Asian fast food restaurant in Tel Aviv. This project consisted of a complete branding process including naming, brand identity, packaging, spatial graphics, as well as the entire planning and interior design of the restaurant itself, including custom-made furniture and products. The challenge of this project was to create a cohesive identity within a very small space, the entire restaurant being the size of approximately 60sqm. The final result was a melange of Asian street market imagery, a bright, bold colour palette, big typography and collage techniques, combined with a high-end atmosphere that fits right into the heart of Tel Aviv.

Design agency: Praktik Photography: Peled Studios and DY Photography

乔普餐厅是特拉维夫的一家亚洲风味快餐厅。项目涉及餐厅的全套品牌设计，包括命名、品牌形象、包装、空间图形、整体规划以及餐厅的室内设计（包含定制家具和产品）。项目所面临的挑战是打造一个能适应小空间的整体形象，因为整个餐厅的总面积只有约 60 平方米。最终的设计融合了亚洲街头小吃店的形象、大胆明亮的色彩搭配、大字体和拼贴画技术，餐厅的高端氛围与特拉维夫市中心的地理位置十分相配。

设计机构：Praktik 设计公司 摄影：Peled 工作室、DY 摄影

SEOUL, KOREA

FISHCAKE

Simplicity Is Ultimate

鱼饼店 / 韩国，首尔

简约，至上

Fish n Cake is a fish cake store located in Seoul, South Korea. Fish cake is a common street food along with Topokki (Korean snack food made from soft rice cake and sweet spicy sauce). However, Fish n Cake wanted to sell fish cake as a side dish of a meal and for quality gift. Eggplant Factory designed various packages for Fish n Cake while maintaining their current brand identity. The designers suggested client to sell heart-warming take-out fishcake contained with take-out coffee cup. Oriented strand boards were used throughout the interior of the store.

Design agency: Eggplant Factory Creative director: Jeyoun, Lee Designers: Hanme, Choi / Jaeseung, Shim / Bobae, Kim / Youngji, Jung Photography: Jaeseung, Shim Client: Fish n Cake

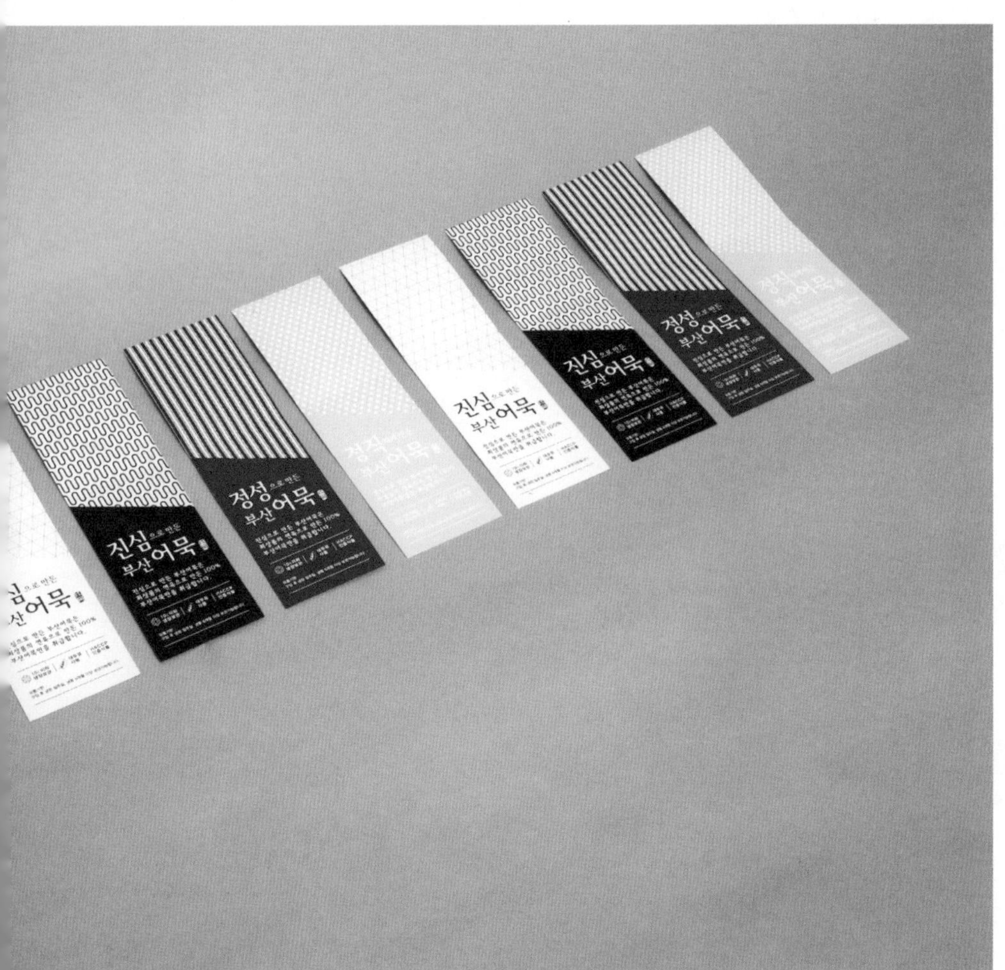

鱼饼店是韩国首尔一家专营鱼饼的餐厅。鱼饼是韩国一种常见的街头小吃，常与炒年糕一起搭配。但是，这家鱼饼店希望将鱼饼作为一道配菜和高品质礼品进行销售。在保持现有品牌形象的前提下，Eggplant Factory 为餐厅设计了各种各样的包装。设计师建议餐厅用外卖咖啡杯来销售暖心的外卖鱼饼。餐厅的室内设计大量使用了定向刨花板，给人以清爽干净的感觉。

设计机构：Eggplant Factory 设计公司 创意总监：李俊勇 设计师：蔡汉美、申俊秀、金波贝、郑永基 摄影：申俊秀 委托方：鱼饼店

P-B
TRUST the CRUST

PIZZA HOUSE

比萨店

MOSCOW, RUSSIA

KOODOO

Koodoo and His Team!

弯角羚餐厅 / 俄罗斯，莫斯科

弯角羚和它的团队们！

Koodoo is a new fast casual pizza and wok restaurant recently opened in Belgorod. Koodoo – a small and cozy place, with an unusual combination of pizza and wok menu and open kitchen, where visitors can watch them cooking. Koodoo's main priority is quality and fast delivery, which allows you to bring the pizza and wok to the buyer at its best, without loss of palatability. Therefore, as the main image of the brand, it was decided to use the Koodoo antelope, one of the fastest animals in the world. Koodoo and his team are in constant motion, moving to a variety of modes of transport, always deliver the food on time. Handmade and playful character graphics corporate identity is designed to support the overall concept of the brand, creating a sense of movement.

Designer: Luda Galchenko Client: Food Republic

弯角羚餐厅是一家专营比萨和烧菜的休闲快餐厅。餐厅小而舒适，奇妙地糅合了比萨、炒菜和开放式厨房，让食客们可以观察大厨的烹饪过程。餐厅注重菜品品质和快速送达，让食客们能品味到口味最佳的菜品。因此，餐厅品牌的主要形象选择了弯角羚——世界上跑得最快的动物之一。弯角羚和它的团队处在不间断的运动中，采用各种交通方式保证美食按时送达。由设计师手绘的趣味形象图形支撑着品牌的整体概念，给人以一种运动感。

设计师：卢达·加尔陈科 委托方：Food Republic 公司

BARCELONA, SPAIN

LA VITA IN FIORI

Symbolic Christmas Colours

菲奥里生活餐厅 / 西班牙，巴塞罗那

标志性的圣诞色

Graphic identity for the pizza restaurant: "La Vita in Fiori" in Sant Cugat, Spain.

Design agency: Lo Siento Studio Client: La Vita in Fiori

项目是为西班牙巴塞罗那圣库加特地区的一家比萨餐厅“菲奥里生活餐厅”所提供的品牌形象设计。

设计机构：Lo Siento 工作室 委托方：菲奥里生活餐厅

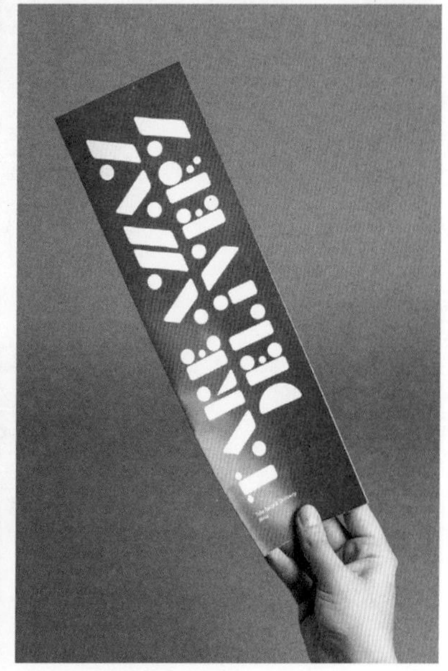

RIO CLARO, BRASIL

NICKS

A Long-range Design for Mr. Nicolau

尼科斯餐厅 / 巴西，里奥克拉鲁

为尼古拉先生的长远目标做设计

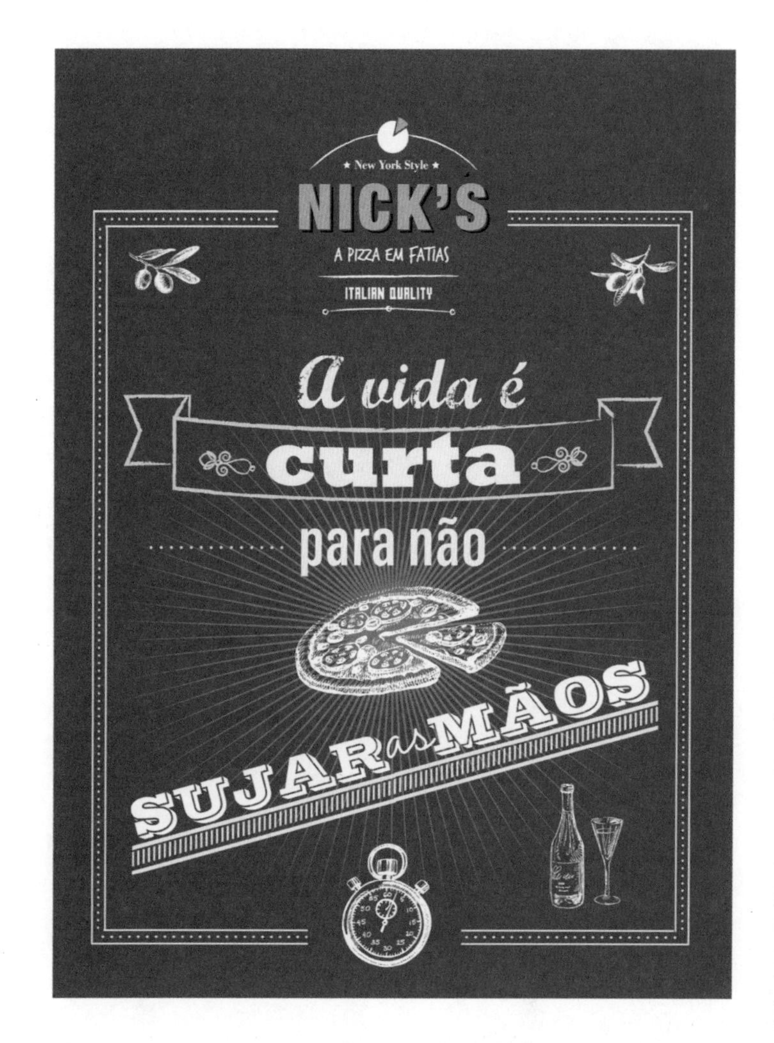

e inspiration came from the name of the brand owner, Nicolau, who hired LOKO
esign to create the whole visual concept of its new business model, a pizzeria inspired
odels Yorkers with Italian recipe where you can eat pizza with your hands. Because
e customer has an expansion plan of shops, create a replicable concept for a future
twork of franchises. Visual communication work with isolated pieces alluding to the
oduct. The designers put the step by step of how to eat pizza, recipes and some
ctures merged together with individual letters that form the word pizza.

sign agency: LOKO Design Client: NICKS

厅的设计灵感来自于它的老板尼古拉，他委托LOKO设计公司为这个新的业模型进行全套的视觉概念设计。尼科斯餐厅是一家比萨饼店，在这里，可以手抓意大利比萨。因为委托人有打造连锁店的目标，因此必须打造一适用于各个地点的品牌概念。视觉传达设计和独立的图形全都暗指餐厅的品。设计师一步步地展示了如何食用比萨饼，绘制了菜单并通过图形与字的融合拼出了“PIZZA”（比萨）这个单词。

计机构：LOKO设计公司 委托方：尼科斯餐厅

Pizza
Barboni
Wood Fire
TRUST
the
CRUST
Pizza
Barboni
Pizza
Barboni
Pizza
Barboni
Fresco!

REAL
BEARDS
REAL
PIZZA
REAL
Fresh!
ITALIAN MADE
P-B
TRUST the CRUST
WOOD
FIRE
Goodness
DELIVERED!
-X-
ITALIAN MADE
PIZZA
BARBONI

JAKARTA, INDONESIA

PIZZA BARBONI

Vintage Style in Street Illustrations

巴博尼比萨 / 印度尼西亚，雅加达

街头插画的复古情怀

Brand Identity and Art Direction created for a local Italian pizza joint in Jakarta City Indonesia.

Design agency: Ageless Galaxy Photography: Kromka Client: Pizza Barboni

项目是为印度尼西亚雅加达市的一家意大利比萨店所提供的品牌形象设计和艺术指导。

设计机构：Ageless Galaxy 设计公司 摄影：Kromka 委托方：巴博尼比萨

RUTA DE
EVACUACIÓN

BARBECUE RESTAURANT
烧烤店

The owner of "La Fama", wanted to bring the original southern USA bbq experience to Colombia. The name of the restaurant was inspired by the idea that butcheries in Colombia are known as "famas", so this was a perfect concept to describe a place where the "butcher" personally explains to the diners what they are about to eat with their bare hands. The design task at Indice was to recreate the feeling of a local butchery while merging it with the restaurant's unique "homely" concept.

Design agency: Indice Designers: Camila Muñoz, Carlos Beltrán Photography: Carlos Beltrán Client: Grupo Artak

肉店烤肉餐厅的老板希望把最正宗的美国南部烤肉带到哥伦比亚，餐厅的名字“LA FAMA”来自于哥伦比亚语中的“肉店”。在餐厅里，会有专业的“屠户”向食客解释他们即将品尝的是那一块肉。Indice 设计公司的任务是重塑一种当地肉店的感觉，同时又要融入餐厅独特的“居家理念”。

设计机构: Indice 设计公司 设计师: 卡米拉·穆诺兹、卡洛斯·贝尔川 摄影: 卡洛斯·贝尔川 委托方: Artak 集团

BOGOTA, COLOMBIA

LA FAMA BARBECUE

Perfect Combination of Bogota Local Butchery and "Homely" Concept

肉店烤肉餐厅 / 哥伦比亚，波哥大

波哥大本土肉店风格与"居家理念"的完美混搭

The Streetz identity is composed of word mark applied to signage, and brought to life in environments created with an edited colour palette, unique typography and messaging to add personality. The primary identity and signage is reminiscent of signs and ephemera from vintage American dining establishments, from hot dog stands to burger joints. These elements work together to create an experience that is sleek and approachable. Similarly, the restaurant interiors use a palette of colours, materials and architectural details to support a brand that is unique, inviting, and scaleable.

Design agency: Cue Creative director: Alan Colvin Designer: Laura Belle Wright

街头餐厅的品牌形象由文字构成，精心编排的色彩搭配、独特的字体和信息内容为餐厅增添了个性。餐厅的主要形象和标识参考了复古美式餐饮店（如热狗店、汉堡店）的设计。这些元素共同营造出一种时尚而不失亲切的体验。同样的，餐厅的室内设计也通过色彩、材料和建筑细节的搭配支撑了这个独特、友好、伸缩自如的餐厅品牌。

设计机构：Cue 设计公司 创意总监：阿兰·科尔文 设计师：劳拉·贝尔·怀特

HOPKINS, USA

STREETZ AMERICAN GRILL

A Fashion and Friendly Feeling Created by American Vintage Style

街头美式烧烤餐厅 / 美国，霍普金斯

美式复古风格营造时尚而又亲切的感觉

SEATTLE, USA

MILLER'S GUILD

Visual Impression of Icons

磨坊主烤肉餐厅 / 美国，西雅图

图标的视觉印象

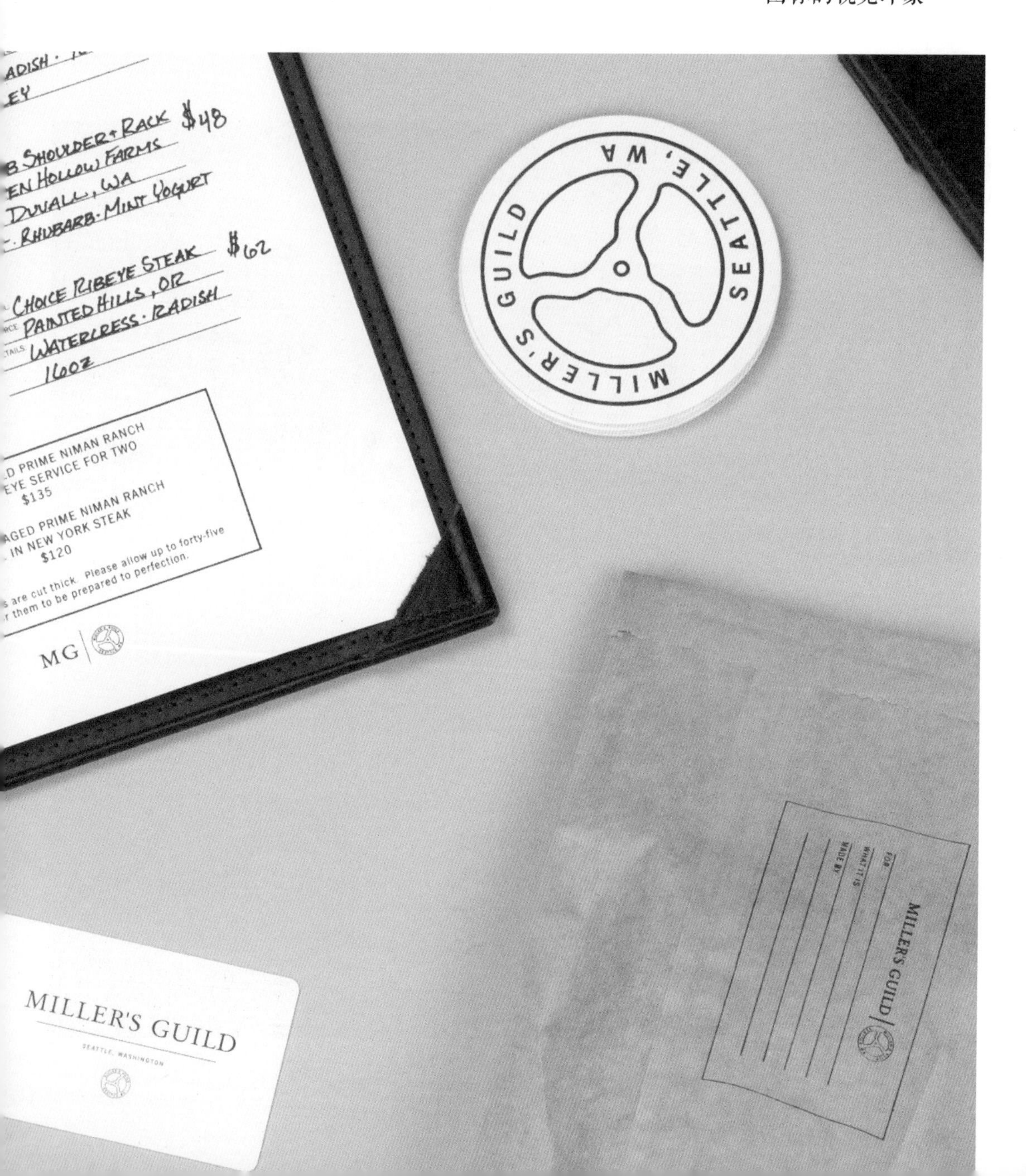

Miller's Guild is a wood-fire grill restaurant Seattle, Washington started by James Be Award-Winning Chef Jason Wilson. The cr wheel icon was designed to emphasise visually impressive grill in an understated c elevated mark. Contrasting the clean li and curves of the identity with the Che handwriting created a way to highli Wilson's individual artistry and persona Pairing these elements with the wheel c upscale classic physical elements, like d woods and leather bound menus, resulted i system of unfamiliar familiarity.

Design agency: Public-Library Designe Ramón Coronado and Marshall R Photography: Rina Jordan, Jason Ware Cli Miller's Guild

磨坊主烤肉餐厅是西雅图的一家炭火烤肉餐厅，由詹姆斯·彼尔德美食大奖获得者詹森·威尔森创立。曲柄轮图标的设计突出了烤架给人的视觉印象。图标简洁的曲线与主厨的手写文字形成对比，突出了主厨的个人技巧和个性。这些元素与车轮图标以及深色木材、皮面菜单等高档元素相融合，形成了一种熟悉而又陌生的感觉。

设计机构：Public-Library 设计公司 设计师：拉蒙·科罗纳多、马希尔·雷克 摄影：雷娜·乔丹、詹森·威尔 委托方：磨坊主烤肉餐厅

OTHER CREATIVE
RESTAURANT
其他创意餐厅

VILNIUS, LITHUANIA

STEBUKLAI

An Incredible Baltic Cuisine Restaurant with Wondrous Design and Adventurous Food

奇迹餐厅 / 立陶宛，维尔纽斯

从奇妙变幻的设计到独特大胆的美食，这是一家不可思议的波罗的海美食餐厅

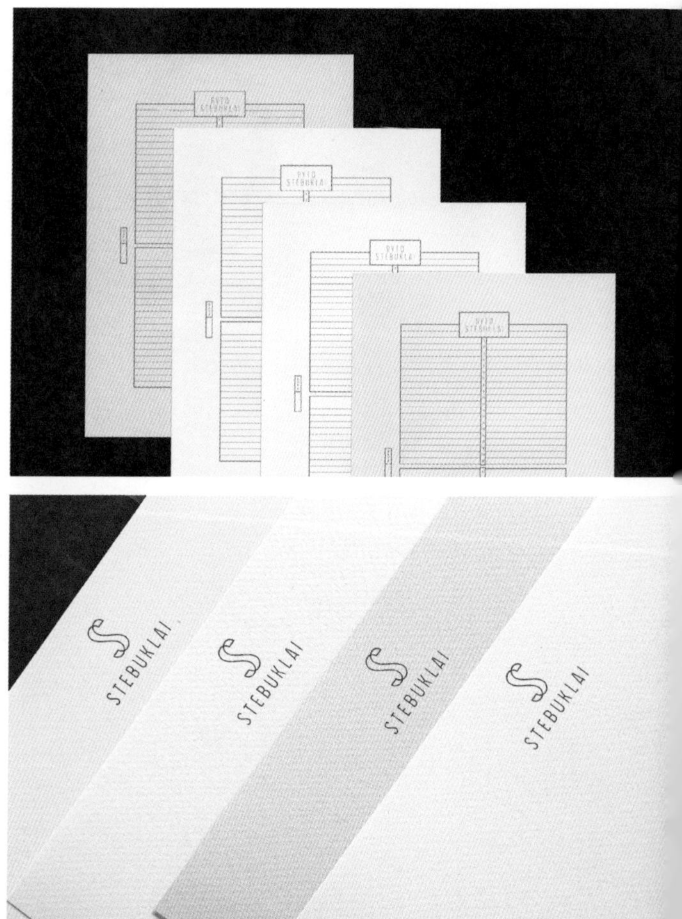

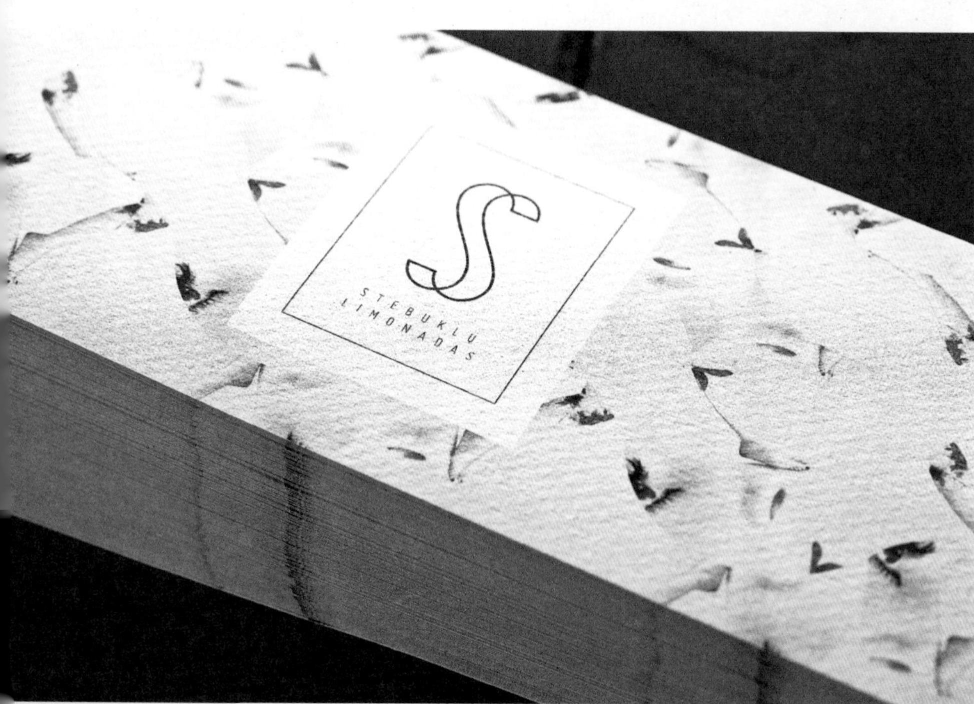

CHALLENGE. To create name, the look, visual identity, even interior for a new Vilnius restaurant specialising in modern Baltic cuisine by one of the young emerging chefs of the region. The food is playful and ever surprising, so the designers needed to think of a name with a hint of hocus-pocus also.

SOLUTION. It will take years for Vilnius to become new Copenhagen, but let's build on the ambition of chef and his team. The restaurant is named "Wonders" - they start with a special, adventurous food and continue throughout the whole experience of dinning. Every little piece, starting from shape shifting logo and menu full of stories and wondrous visuals - serves the central concept of small bits of magic in every step. Or in every bite, rather.

Design agency: New Agency Designer: Migle Rudaityte, New Agency Photography: Robertas Daskevicius Client: UAB Stebuklai

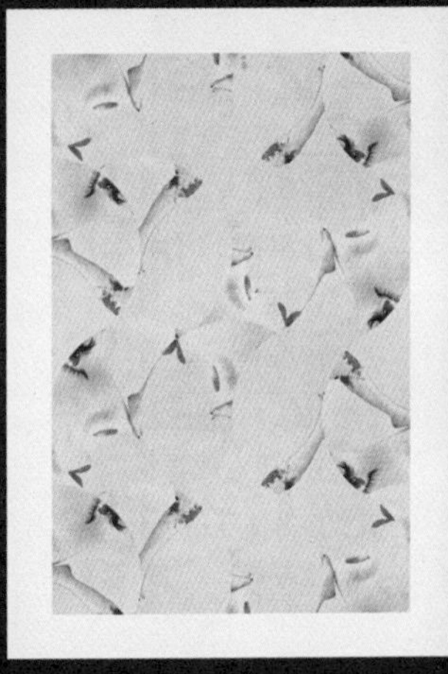
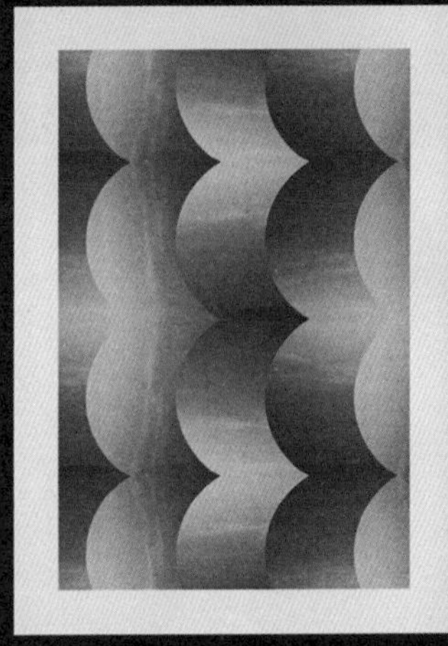

PUSRYČIAI
DARO
STEBUKLUS

STEBUKLAI
DARO
PUSRYČIUS

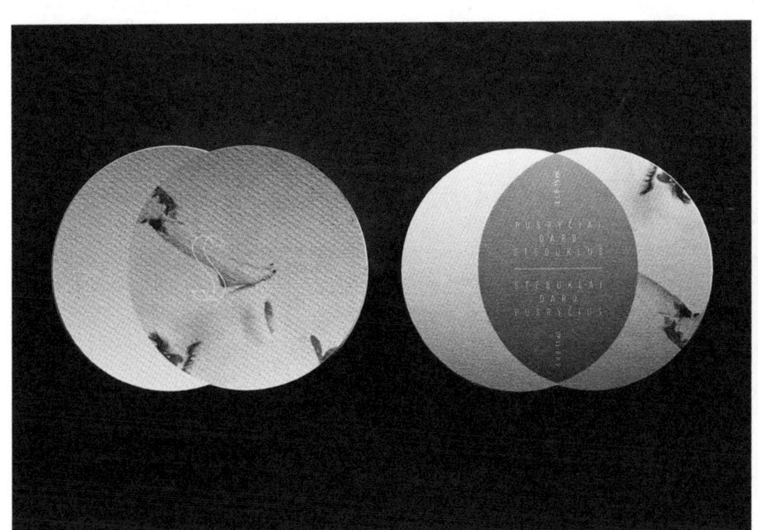

挑战：项目是为一家由新秀主厨经营的现代波罗的海美食餐厅打造全套的命名、外观、视觉形象及室内设计。餐厅的美食花样翻新，因此设计师必须在餐厅的命名中就有所体现。

解决方案：维尔纽斯要想成为新的哥本哈根还需要数年的时间，但是我们仍然相信主厨和他们的团队。餐厅被命名为“奇迹”，以独特、大胆的美食起步，为食客提供全套的就餐体验。每一个细节——从变换形状的品牌 LOGO 到写满故事和奇妙视觉效果的菜单——都给人以不可思议的感觉。当然，奇妙的美食更不在话下。

设计机构：New Agency 设计公司 设计师：米格尔·卢达伊泰特 摄影：罗伯塔斯·达斯科伟修斯 委托方：UAB 奇迹餐厅

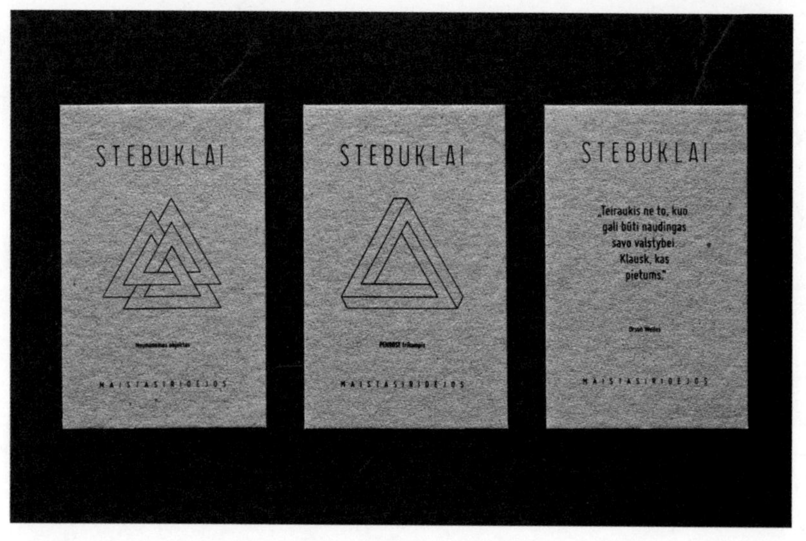

VIENNA, AUSTRIA

"BAKERY"

Independent and Confident Red

"面包房" / 奥地利，维也纳

独立而自信的红

The "Bakery"– lobby, breakfast room and restaurant of Hotel Daniel Vienna in one, distinguishes itself from the base black and white colour scheme of the hotel's logo with a fresh, red variation. Independent and confident, as a magnet for guests and city dwellers, the "Bakery" provides optimal conditions for the transition between check-in, business meetings and relaxed get-togethers with friends.

Design agency: moodley brand identity Designer: Sabine Kernbichler Photography: Marion Luttenberger Client: Weitzer Hotels Betriebsgesellschaft m.b.H.

KEINE SCHLECHTE LAUNE
NO BAD TEMPER

KRAWATTE? PROBIER'S MAL OHNE
TIE? WHY NOT TRY WITHOUT

KEINE BETRIEBSSPIONAGE
NO SPIES

flotte
BIENEN
DIE WELT BRAUCHT MEHR
BIENEN
BIENEN-
STOCK-
DUNKEL.
URBAN STAY
DANIEL
SMART LUXURY
HI HONEY!
Süß! MARZIPAN-
HONIG IM GLAS.
OPTIMISTIC FOR THE WORLD
Bikes & Vespas
15,-
5,-
Einfach machen
Von Blumen und Bienen
Bee Station
BUY IN OUR
SHOP!
1886

ALOIS DANIEL
BAKERY
GOOD
MORNING
I FRÜHSTÜCK/BREAKFAST
Bakery Breakfast
Take Away Breakfast
Yummy!

1886
W
OPTIMISTIC FOR THE WORLD
ALOIS DANIEL
BAKERY
HOTEL EST 1886
banana
kipferl
yoghurt
juice
SPEEDY
BREAKFAST

“面包房”是丹尼尔维也纳酒店集大堂、早餐室和餐厅于一身的空间。它以鲜活的红色渐变 LOGO 与黑白基调的酒店 LOGO 区分开来。独立而自信的“面包房”吸引着酒店的宾客和城市居民，为办理入住手续、进行商务会面和亲友放松小聚的人们提供了绝佳的休闲空间。

设计机构：moodley 品牌形象设计公司 设计师：萨宾·科恩比切勒 摄影：马里恩·卢顿伯格 委托方：威兹酒店集团

POZNAN, POLAND

KOMBINAT

LOGO: Not a Purely Mix of Typefaces

联合餐厅 / 波兰，波兹南

LOGO: 不是字体特效的堆砌

The designer's task was to design logo for the new restaurant in Poznań, Poland – KOMBINAT, which offers snack dishes, based on healthy, seasonal, Polish cuisine. Logo was about to be as simple as possible, so the illustrations are in the background – building the foundation for the strong typography. Lettering is based on Libel Suit font from Typodermic Fonts. All elements are used in black and white colour only – on black, white or natural-brown background.

Designer: Maria Mileńko Photography: KOMBINAT Client: KOMBINAT

设计师的任务是为这家位于波兰波兹南的新餐厅设计一个LOGO。联合餐厅主营健康的波兰时令小吃。LOGO的设计越简单越好，因此设计师在LOGO的背景加入了插画，为有力的排版设计奠定了基础。设计的字体以Libel Suit字体为主。所有元素全部采用简单的黑白两色，背景为黑、白或天然的棕色。

设计师：玛利亚·米兰可 摄影：联合餐厅 委托方：联合餐厅

KOMBINAT
BROWN SUGAR
WHITE SUGAR
TODAY'S MENU BY
KOMBINAT
spinach-mushroom cutlets
dumpling with chine
kale in tempura
tomato cream soup
dumpling with cured
WATERMELON SODA BY
SANDWICH WITH MACKEREL BY
HERBAL TEA BY
HERBS MIX BY
PURE BLACK COFFEE BY
SPICY MIX BY
SMALL COFFEE BY
MEDIUM COFFEE BY
BIG COFFEE BY

BOGOTÁ, COLOMBIA

BLACK BEAR

Vintage Design Reminds You of Your Grandparents' House

黑熊餐厅 / 哥伦比亚，波哥大

复古设计雕刻出如爷爷家般的温暖时光

Black Bear is a special place with a story behind its name: it is how chef Andrew Blackburn's closest friends call him. This name and the restaurant's architecture were the source of inspiration for the development of its identity. Black Bear does not belong to a specific time period, although it does remind you of your grandparents' house. This can be observed in its classic typography, its sober layout, its copper signage and its leather menus. The bear is the central character whose noble and expert face invites us to a place dedicated to our enjoinment and pleasure.

Design agency: Indice Designers: Camila Muñoz, Paulina Carrizosa Photography: Esteban Rodriguez Client: Grupo Takami

黑熊餐厅的名字背后有一段故事：“黑熊”正是主厨安德鲁·布莱克本的昵称。餐厅的名字和建筑设计都是品牌形象开发的灵感来源。黑熊餐厅不属于任何一个特定的时代，但是它能让你想起爷爷的家。古典的文字设计、冷静的布局、铜招牌和皮质菜单，这一切都具有浓浓的复古感。黑熊无疑是设计的核心，它用高贵而专业的面庞吸引着我们来享用美食。

设计机构：Indice 设计公司 设计师：卡米拉·穆诺兹、宝林娜·卡里索萨 摄影：埃斯特班·罗德里格斯 委托方：高美集团

Black Bear
COCINA Y RAW BAR
Black Bear
WWW.BLACKBEAR.COM.CO
PRICELESS

Cocina
Calidad
Presentación
Variedad
Sabor
Ambiente
Decoración
Música
Temperatura
¿Por qué?

Black Bear
COCTELES BLACK BEAR
BOGOTA MULE. $ 17.800
RON AÑEJO, JENGIBRE FRESCO, SODA.
GIMLET DE PEPINO. $ 21.800
GINEBRA, PEPINO Y LIMON
CARBONATED MARGARITA. $ 21.800
TEQUILA BLANCO, LIMON, AGAVE Y BITTERS DE NARANJA
MAPLE DERBY. $ 19.800
BOURBON, TORONJA, LIMON, MAPLE
DAIQUIRI. $ 17.800
PENICILLIN.
SCOTCH, GINGER, MIEL Y LIMON
LONSDALE. $ 21.800
GINEBRA, LIMON, JUGO DE MANZANA, MIEL, ALBAHACA
AGAVE OLD FASHIONED. $ 21.800
TEQUILA REPOSADO, NECTAR DE AGAVE, BITTERS DE CHOCOLATE Y TORONJA
KENTUCKY 75. $ 18.800
BOURBON, MELOCOTON, MENTA Y PROSECCO
CLOVER CLUB. $ 21.800
GINEBRA, DRY VERMOUTH, FRAMBUESA Y CLARA DE HUEVO
RANGOON GIN COBBLER. $ 22.800
GINEBRA, COINTREAU, AGAVE, BITTERS DE NARANJA Y ANGOSTURA

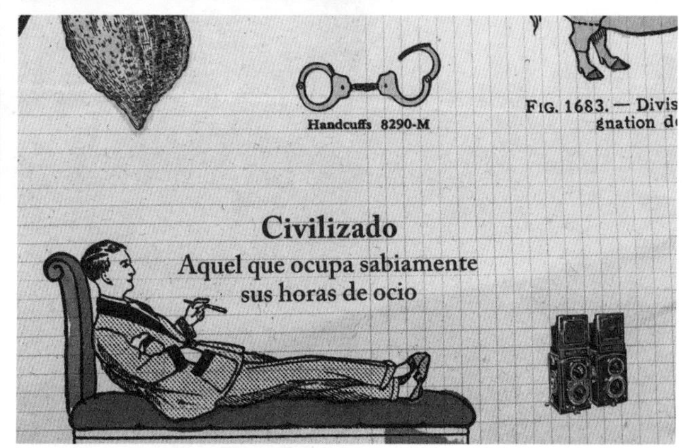

2013 年初，主厨劳拉·隆多尼奥从澳大利亚来到了哥伦比亚，她梦想开一家自己的餐厅。她找到了 Indice 设计公司，与他们分享了自己打造一个创新的休闲餐吧的想法。餐厅的百科全书式图形设计以主厨最近的烹饪求学之旅为灵感，体现了意大利、法国、越南、泰国和澳大利亚的风土人情。

设计机构：Indice 设计公司 设计师：卡米拉·穆诺兹、宝林娜·卡里索萨 摄影：埃斯特班·罗德里格斯 委托方：劳拉·隆多尼奥

MEDELLIN, COLOMBIA

OCIO RESTAURANT

Head Chef's Educational Culinary Travels

奥西奥餐厅 / 哥伦比亚，麦德林

主厨的烹饪求学之旅

In early 2013, chef Laura Londoño came from Australia to Colombia with the dream of opening her own restaurant. She then looked for Indice and shared her idea a creative, informal restaurant and bar which she would call Ocio. The encyclopaedic imagery was inspired by head chef recent educational culinary travels which included Italy, France, Vietnam, Thailand and Australia.

Design agency: Indice Designers: Camila Muñoz, Paulina Carrizosa
Photography: Esteban Rodriguez Client: Laura Londoño

BERKSHIRE, UK

THE BARN

A Relaxed Atmosphere Created by Healing Illustrations

谷仓餐厅 / 英国，伯克郡

治愈系插画带出轻松氛围

As part of the branding job & SMITH did for Coworth Park (a Dorchester Collection hotel), they created this identity for their informal dining restaurant. The designers wanted customers to feel that The Barn didn't take itself too seriously – a good contrast to the main hotel.

Design agency: & SMITH Designer: Sam Kang
Photography: & SMITH

项目是 & SMITH 设计公司为科沃斯庄园酒店（一家多切斯特精品连锁酒店）所提供的品牌形象设计的一部分，这是酒店内部的一家休闲餐厅。设计师希望消费者感受到谷仓餐厅那种随性的氛围，使其与酒店的主品牌形成对比。

设计机构：& SMITH 设计公司 设计师：山姆 · 康
摄影：& SMITH 设计公司

MEDELLIN, COLOMBIA

SUEGRA SABORES CASEROS

Warmth through Bright Colours and Wood

婆婆餐厅 / 哥伦比亚，麦德林

明亮色彩与木质相契的温暖

Suegra is a tribute to the homemade food with which we all grew up, reinterpreted by talented Chef Juan Pablo Valencia, who transports us to other times and places through your unique gastronomic offer.

Design agency: Mantra Branding Designers: Wilson Aristizabal, Andrés Agudelo Client: Cocina en Evolución

婆婆餐厅专注于提供我们从小到大吃的家常自制美食，优秀的主厨胡安·巴勃罗·瓦伦西亚通过独特的菜品将我们带到了另一个时空之中。

设计机构：Mantra 品牌设计公司 设计师：威尔逊·阿里斯蒂萨鲍尔、安德烈·阿古德洛 委托方：Cocina en Evoluci ó n 餐饮公司

SIEMPRE HAY UN
LUGAR EN NUESTRA
Mesa
PARA DAR
Y
CONVIDAR
Tradición
¡Aquí todo es Rico!
MÁS AMOR,
MÁS SABOR
ESPONTÁNEO
ALE-
GRÍA
Cariño
COMO EN
CASA
Suegra
sabores caseros
GENIAL
COTIDIANO
BARRIGA LLENA
CONTENTO
DONDE COME 1
COMEN TODOS
Come, TOMA, Pica
¡QUÉ DELICIA!
SABORES
caseros
Calor de
HOGAR
SONRISAS
SORBA TRANQUILO
QUE ESTÁ EN SUEGRA
alegría
colombiana

Come, TOMA, Pica Cariño
Aderezos
Salsa de
AJÍ
Aderezos
Vinagreta
Light
Aderezos
Salsa
BARBECUE
PICANTE
Suegra
Aderezos
CONVIERTE
LO ORDINARIO
EN EXTRAORDINARIO

POSTRES | deserts
deli

ARRIMATE una SILLITA

Grønbech & Churchill is a gourmet restaurant in Copenhagen. Chef Rasmus Grønbech's culinary expression arises from the idea of delicately refining raw produce with simple, yet skillful intervention. The identity takes its inspiration from this culinary process. The transformation from raw to refined is expressed throughout the identity by juxtaposing contrasting letter forms, colours and textures.

Design agency: Re-public Designer: Romeo Vidner Photography: Jenny Nordquist Client: Grønbech and Churchill

格伦比赫与丘吉尔餐厅位于哥本哈根，主厨拉斯姆斯·格伦比赫以简单而又巧妙的制作方法将精致的原材料变成可口的菜肴。餐厅的形象设计从这一烹饪过程中获得了灵感，通过字体、色彩和纹理的对比体现了从原料状态到精致菜肴的转变过程。

设计机构：Re-public 设计公司 设计师：罗密欧·韦德纳 摄影：珍妮·诺德齐斯特 委托方：格伦比赫与丘吉尔餐厅

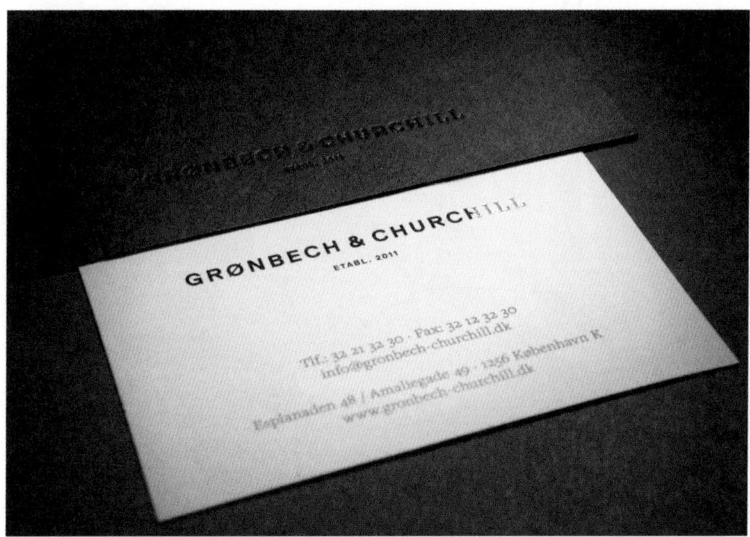

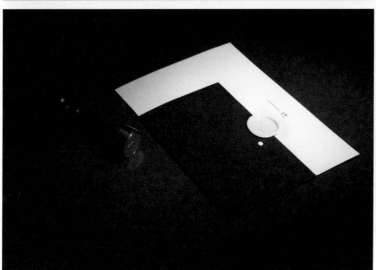

COPENHAGEN, DENMARK

GRONBECH AND CHURCHILL

Great Colour Combination of Black, White and Gold

格伦比赫与丘吉尔餐厅 / 丹麦，哥本哈根

黑白金的色彩气质

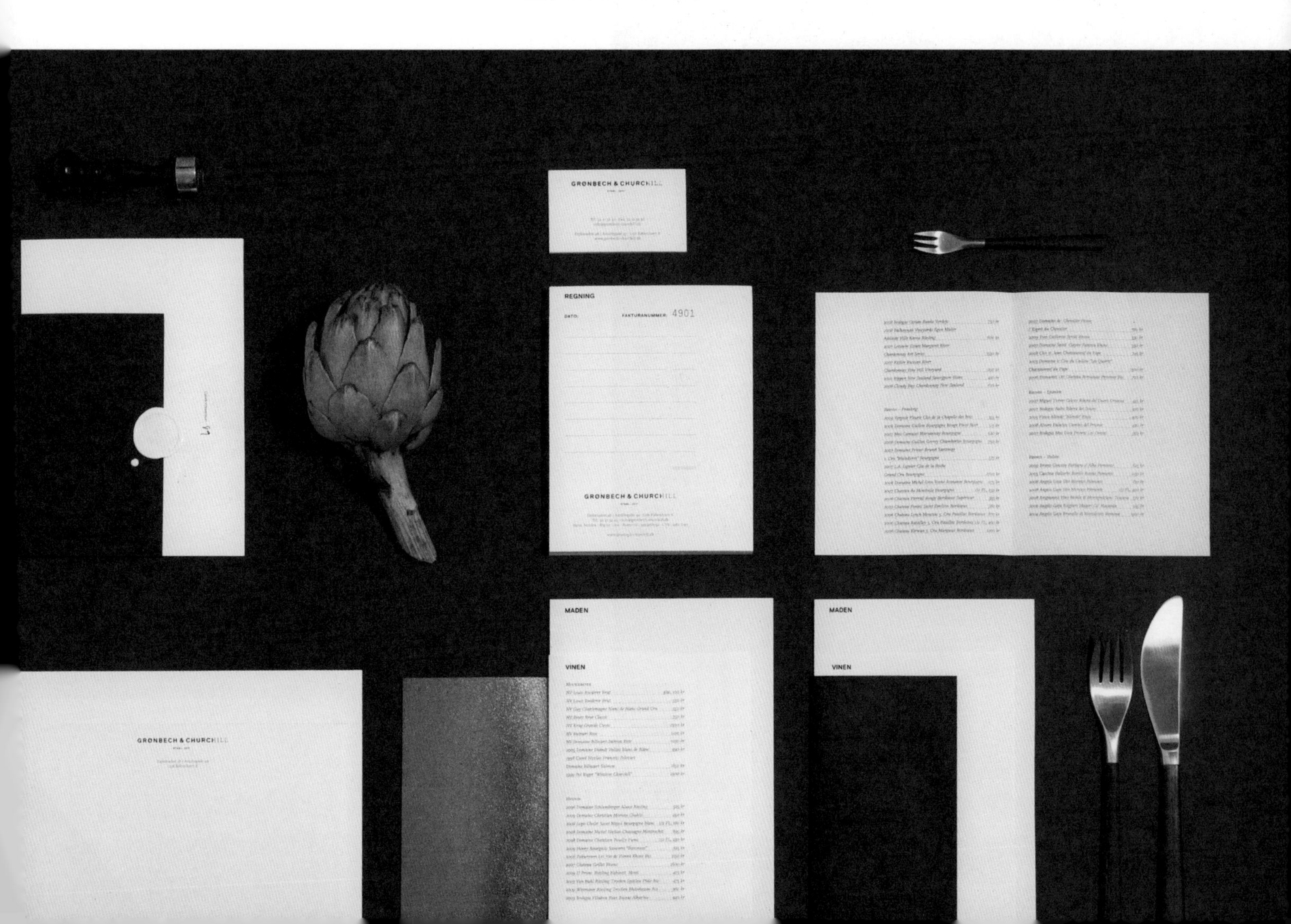

BERGEN, NORWAY

BARE RESTAURANT

Typography: A Movement of Words

贝尔餐厅 / 挪威，卑尔根

版式设计：文字的乐章

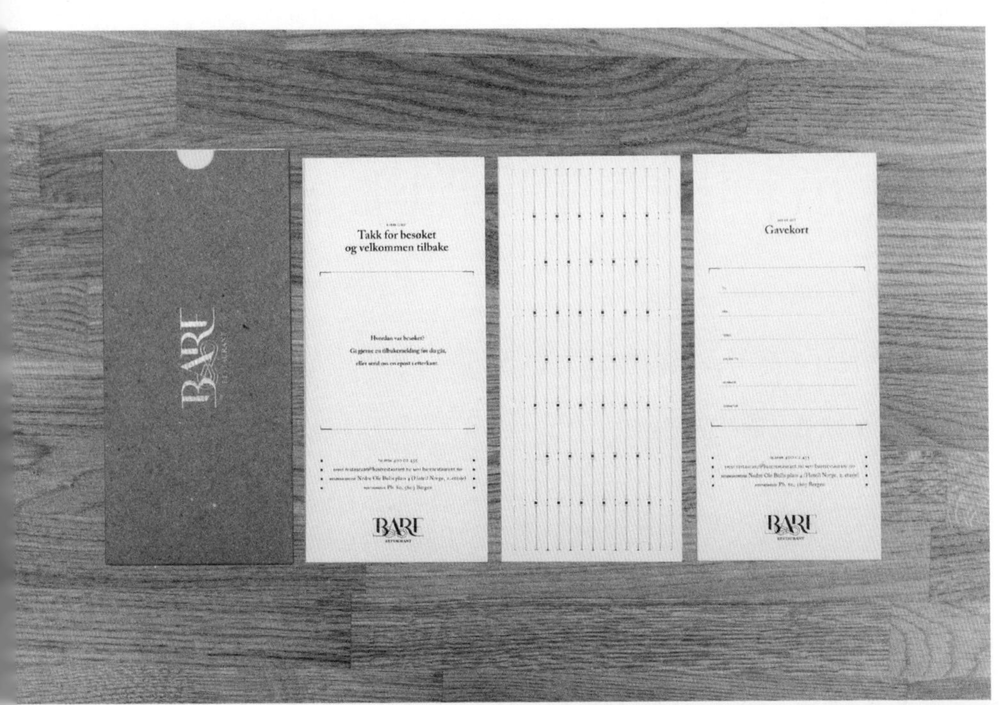

Bare is a high end restaurant with amazing chefs. Haltenbanken has developed a visual identity that shows the exclusive quality and the exquisite flavours you will experience visiting the restaurant. The use of photography as a pillar in the identity shows off the masterpeices Bare's chefs are creating.

Designer: Haltenbanken Photography: Bent Rene Synnevåg

贝尔餐厅是一家拥有优秀主厨的高端餐厅。Haltenbanken设计公司所开发的视觉形象展示了餐厅的卓越品质和细腻口味。照片的运用展示了餐厅主厨的高超技艺和精美菜肴。

设计师：**Haltenbanken设计公司** 摄影：**本特·里尼·辛耐瓦格**

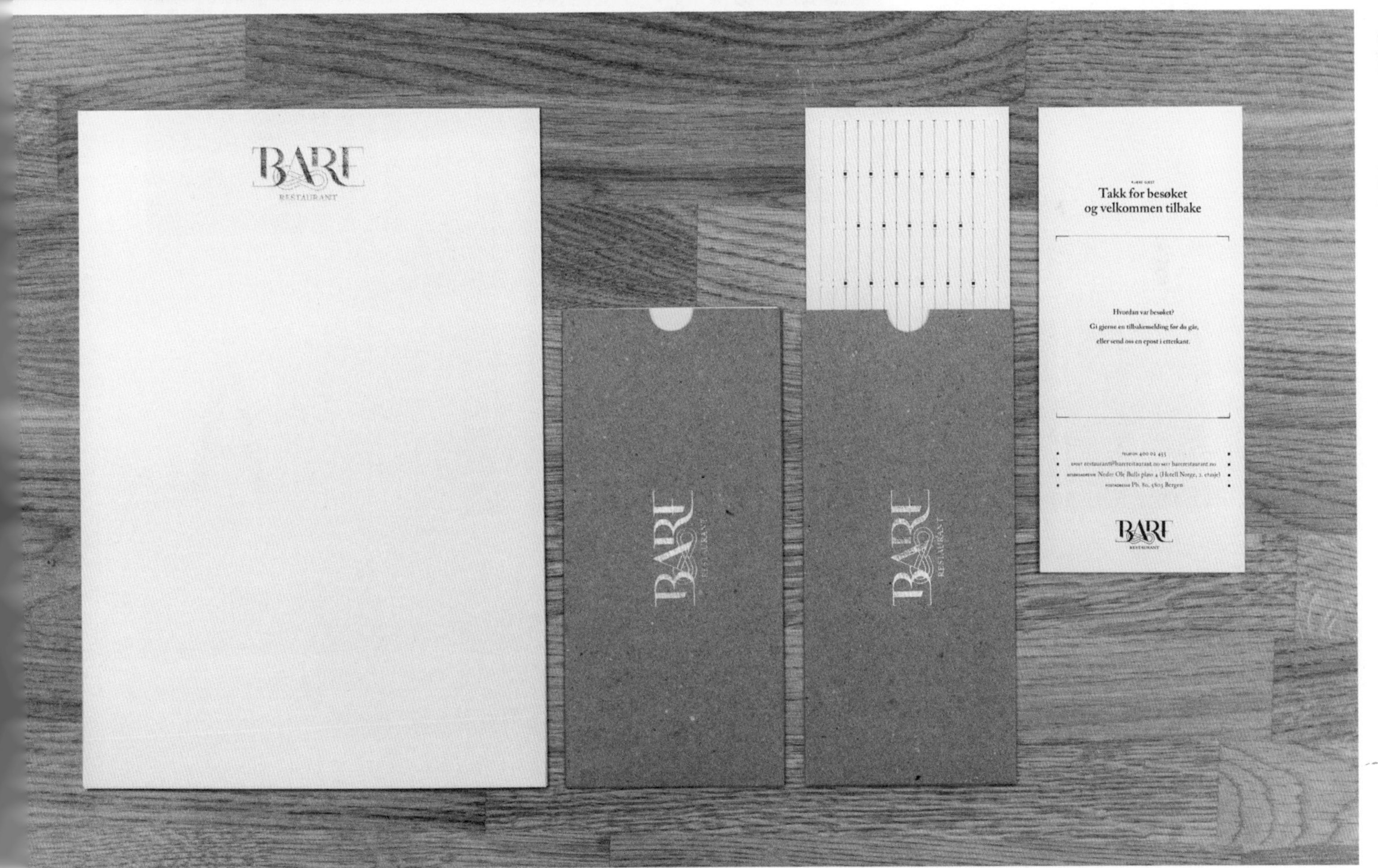

STUTTGART, GERMANY

RESTAURANT BERG

Symbolic Imprint of a Wine Glass

伯格餐厅 / 德国，斯图加特

象征意义的红酒杯印

The stylised imprint of a wine glass enfolds the geometrical majuscule Letters of the Restaurants Name "BERG", a reference to the owners reputation as a sommelier as well as to the focus of the Restaurant. Main colours are a bright, vivid orange and a distinguished cool gray. Business cards and 500 invitations are refined with hand embossed gold foil stickers. Their envelopes are sealed with the emblem on a golden sealing wax. Menu and the list of beverages have a blind embossing on their wrapping which is made of Toile-du-Marais-Linen.

Design agency: ADDA Studio Creative director: Christian Vögtlin Photography: ADDA / Stephanie Trenz Client: Restaurant Berg

12. & 13.05.12
RESTAURANT BERG
AB 17:30 UHR

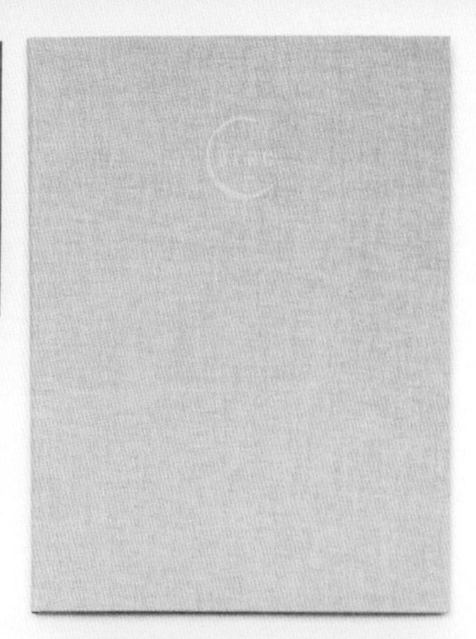

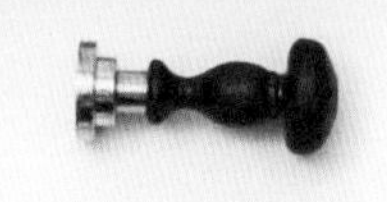

BERG

PHILIPP BERG

JOHANNES DAIKER

RESTAURANT BERG
JOB

15.05.2012
ERÖFFNUNG
RESTAURANT BERG

BERG

LIEBE GÄSTE,
LIEBE FREUNDE,
ZUM 1-JÄHRIGEN BESTEHEN MÖCHTE ICH MICH BESONDERS BEI IHNEN FÜR DEN ERFOLG DES RESTAURANTS BEDANKEN
ZWISCHEN DEM 21. & 25.05. LADE ICH SIE HERZLICH DAZU EIN, DIE HÖHEPUNKTE DES VERGANGENEN JAHRES NOCH EINMAL REVUE PASSIEREN ZU LASSEN.
ICH FREUE MICH AUF IHREN BESUCH
IHR PHILIPP BERG & TEAM
À
À LA CARTE

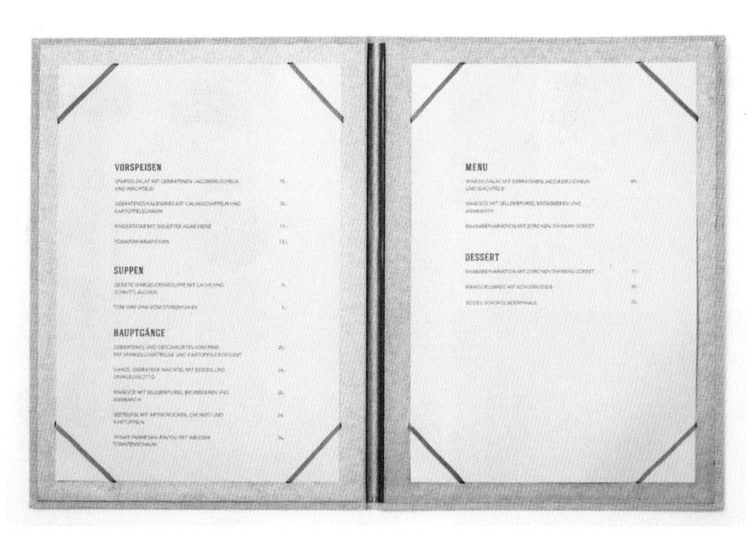
VORSPEISEN
SUPPEN
HAUPTGÄNGE
MENU
DESSERT

MIT PROFESSION &
LEIDENSCHAFT
WÜRTTEMBERG
RIOJA
ELSASS

风格化的红酒杯印记环绕着餐厅名字“BERG”（伯格）的大写印刷体字母，既暗示了餐厅老板的红酒师身份，又突出了餐厅的重点。设计的主色调是明亮活泼的橙色和高雅清爽的灰色。餐厅的名片和 500 张邀请卡全部配有手工粘贴的金箔浮雕贴纸。餐厅的信封采用金色封蜡以纹章密封。菜单和酒水单的薄麻布外壳上带有浮雕印花。

设计机构：ADDA 工作室 创意总监：克里斯蒂安沃格特林 摄影：ADDA/ 斯蒂芬妮·特兰兹 委托方：伯格餐厅

SHEN ZHEN, CHINA

MEETOWN

"Castles, Keys, Old Trees, Animals, Stars, Gardens"

谜堂餐厅 / 中国，深圳

"城堡，钥匙，老树，动物，星空，花园"

As a restaurant and a clubhouse for parent-child gathering, Meetown hopes to bring a dream and surprising space for children, and thus to promote more interaction and communication between parents and children. Taking stories as a clue to connect parents and children, the designers bring children a voyage of discovery, and put together stories about discovery, about bravery, about happiness, and about growth, through the creation of associated elements, castles, keys, old trees, animals, stars, gardens and so on.

Design agency: G'DAY Innovation Designer: JS Chen Photography: JS Chen Client: Meetown

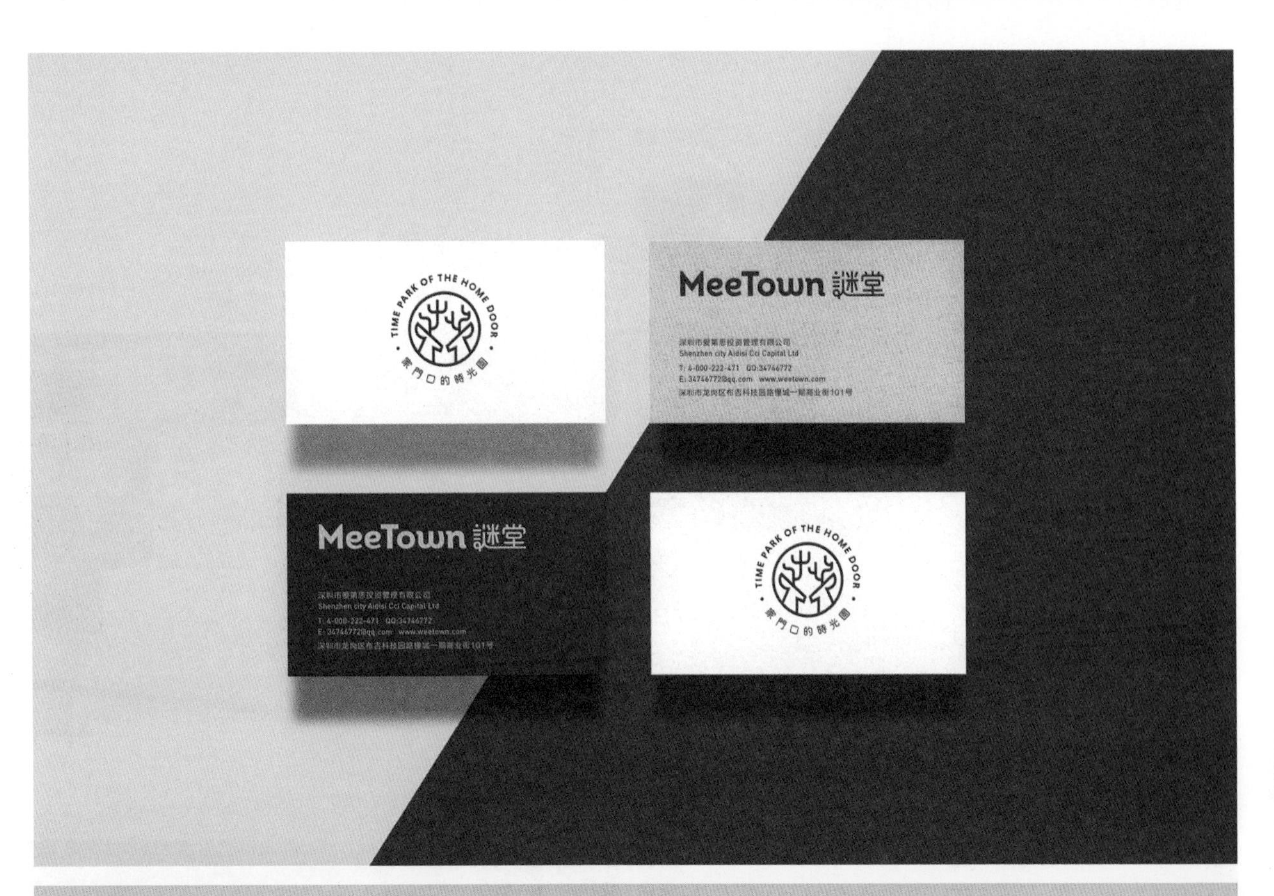

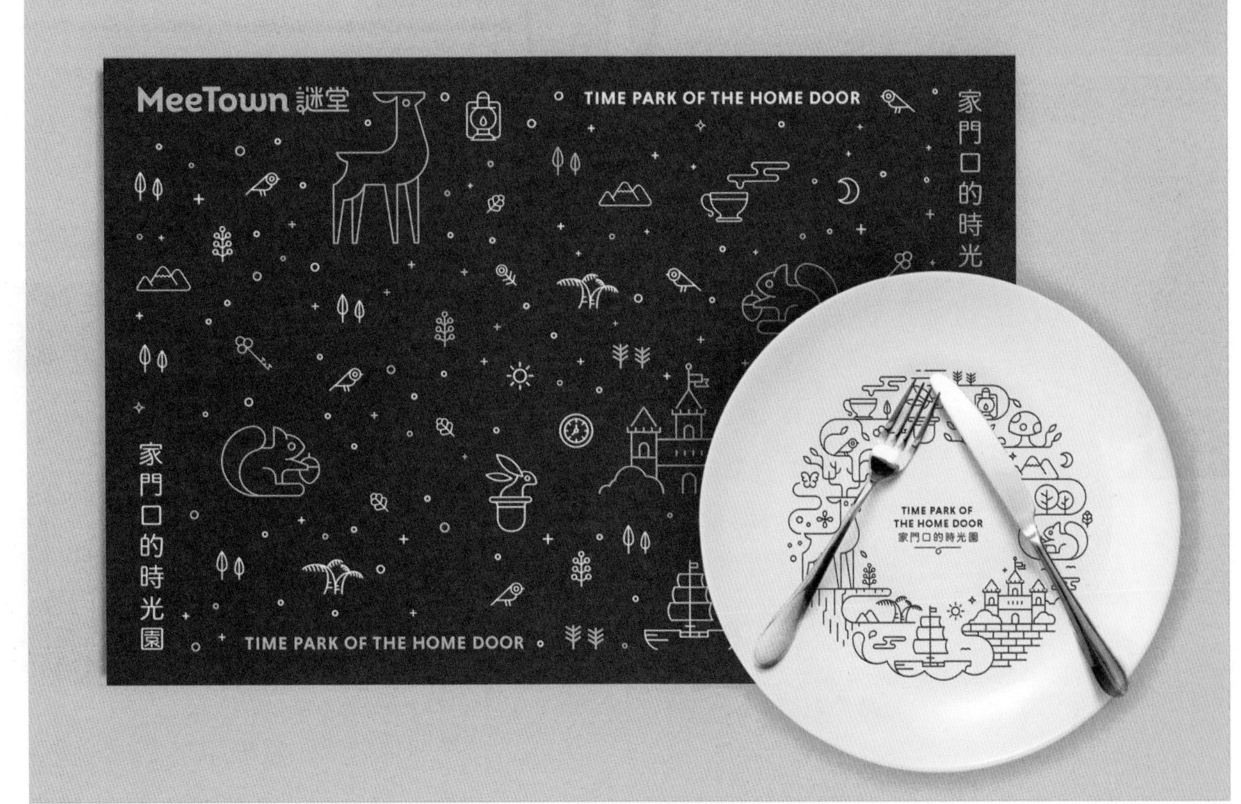

etown 谜堂是一家亲子欢聚餐厅会所，希望带给孩子个有梦想、有惊喜的空间，进家长和孩子更多的互动沟通理解，设计师以故事为连接家长与孩子之间的素，给孩子一个发现之旅，过创建有关联的元素：城、钥匙，老树，动物，星空，园等串起关于探险关于勇关于快乐关于成长的故事。

计机构：G’ DAY 设计公司
计师：陈佳生 摄影：陈佳
委托方：谜堂餐厅

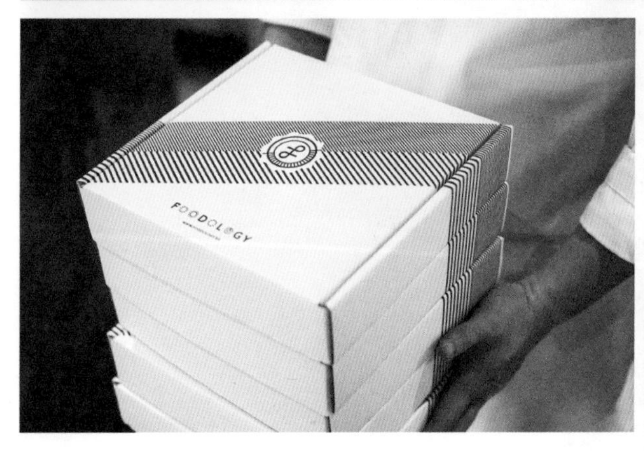

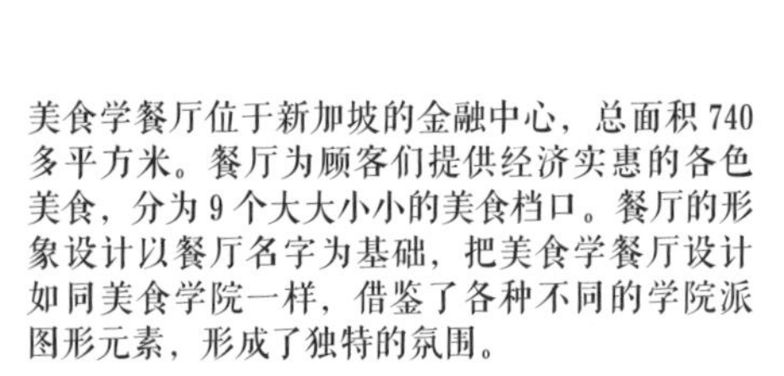

美食学餐厅位于新加坡的金融中心，总面积 740 多平方米。餐厅为顾客们提供经济实惠的各色美食，分为 9 个大大小小的美食档口。餐厅的形象设计以餐厅名字为基础，把美食学餐厅设计如同美食学院一样，借鉴了各种不同的学院派图形元素，形成了独特的氛围。

设计机构：Somewhere Else 设计公司 设计师：蔡梅 摄影：菲利克斯·李 委托方：美食学餐厅

SINGAPORE

FOODOLOGY

Visual Features of Pattern Design

美食学餐厅 / 新加坡

图案设计的视觉特征

Foodology is an 8000sqft restaurant situated in the midst of Singapore's financial district. It serves restaurant grade food at affordable prices via 9 different food types / stations with plans to scale larger or smaller depending on their upcoming locations. Based on the name, the identity treats Foodology as an institution for food and borrows different graphic elements from academia to create its own unique voice.

Design agency: Somewhere Else Designer: May Chua Photography: Felix Lee Client: Foodology One

SINGAPORE

PIDGIN

A Ferryman Who Travels between Now and Past

融合餐厅 / 新加坡

现代与怀旧的摆渡者

Delightful, curious dishes imagined and inspired by food from the streets, from dreams, from travel and adventure. To express this "hodgepodge of ideas", the identity utilises a vernacular design language that combines and mixes a variety of different graphic styles. The overall design much like the food at Pidgin portrays an identity that is contemporary with tinges of nostalgia.

Design agency: Somewhere Else Designer: Madeleine Poh
Photography: Felix Lee Client: Pamplemousse

餐厅的美食令人欣喜而好奇，它们的制作灵感来自于街头小吃、梦境中、旅途上。为了表现这种“思想融合”的概念，餐厅的形象设计采用了本土设计语言和各种图形风格相混合的设计方式。整体设计就像融合餐厅所供应的美食一样，现代而又怀旧。

设计机构：Somewhere Else 设计公司 设计师：玛德琳·波赫 摄影：菲利克斯·李 委托方：柚子餐饮公司

BUSY BUT WE ARE
OPEN
MON TO SAT
SATURDAY
TIPS APPRECIATED
WELCOME · HAVE-A-NICE-DAY
PAMPLEMOUSSE
99¢
ON SALE
FIGS; ARTICHOKES & LOCALLY SOURCED VEGETABLES FRESH CAI XIN
QUALITY
motto
COURTESY
ARTISANAL
BAR
BEERS AVAILABLE
PIDGIN
KITCHEN & BAR
PIDGIN KITCHEN & BAR
7 DEMPSEY ROAD 01-04 SINGAPORE 249671
T: +65 6475 0080 / WWW.PIDGIN.SG
STANDARD
SOUP
OF THE DAY
IMPERIAL PALE ALE
SALMON
SEABASS
CAI FAN
GAROUPA
SCALLOP
PING PONG
BILLIARDS
MAHJONG
家常小吃
NO ORDER TOO BIG
DELIGHTFUL · CURIOUS · DISHES

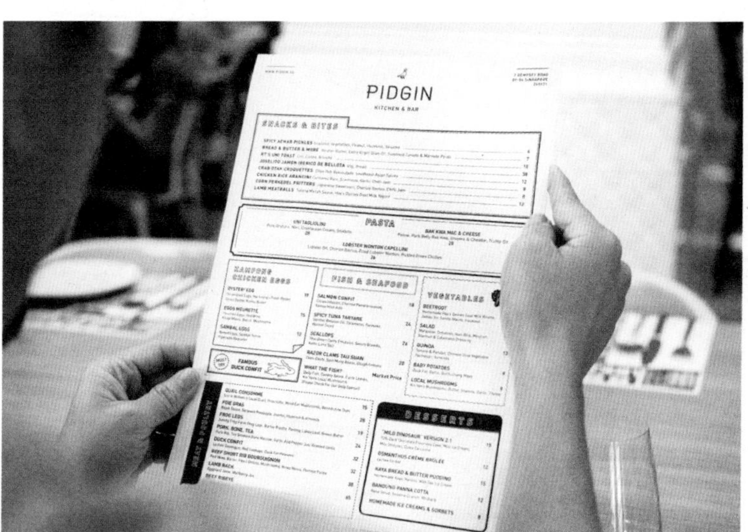
PIDGIN
KITCHEN & BAR
PASTA
DESSERTS

PIDGIN

PUBLIC
SCARMOZA
KAMPONG
CHICKEN
LOWER FLOOR W.C.
CHORIZO IBERICO

SINGAPORE

PODI THE FOOD ORCHARD

Inspired by Spices and Herbs

波蒂美食林 / 新加坡

灵感汲取于调料与香草

Podi, the food orchard, is a new all day restaurant that celebrates bold, robust and unique flavours. Podi is the brainchild of Cedele and upholds Cedele's philosophy of advocating buying and making food deliciously, and most importantly, responsibly. Derived from the meaning of Podi in Hindi – a course mixture of ground dry spices and herbs – a simple and modern logo mark with strong colour accents was developed, drawing inspiration from heaped spoonfuls of various spices and herbs mounds. The earthy tones used throughout in colour palette to imagery is applied to reflect Podi's mission of making good, natural and organic food.

Design agency: Bravo Creative director: Edwin Tan Designer: Jasmine Lee
Client: Bakery Depot

Cedele is moving from
Ngee Ann City to Paragon
and reopening as
Bakery Cafe at #01-25A.
Also proudly launching
PODI, the food orchard
at #01-20A in Paragon.
Opening Mid-September 2012
PODI
THE FOOD ORCHARD
www.podifood.com
PODI, the food orchard is a new all day restaurant in Paragon.
Choose from our delicious selection of memorable sharing plates, mains, pastas, risottos and stone-baked pide pies, paired with refreshing cocktails and relaxing teas.
PODI showcases mouthwatering cakes, pies and pastries for dessert sharing and takeaways.
Take your pick at PODI, the food orchard.
www.podifood.com
PODI, the food orchard is a new all day restaurant in Paragon that celebrates bold, robust and unique flavours.
We joyously concoct western and asian dishes that perfectly combine herbs and spices with the freshest, quality natural ingredients.
PODI's menu design faithfully upholds the Cedele ethos of freshness, quality (use of organic unrefined sugar, no transfat), delicious original recipes handmade from scratch, and most importantly, always buying and making food responsibly.
PODI INTRODUCES
Sharing plates & Stone-baked pide pies
PODI SERVES
All day food
PODI COOKS
Memorable dishes
PODI BLENDS
Refreshing drinks
PODI BAKES
Mouthwatering desserts

PODI
THE FOOD ORCHARD
Cedele is moving from Ngee Ann City to Paragon and reopening as Bakery Cafe at #01-25A.
Also proudly launching PODI, the food orchard at #01-20A in Paragon.
Opening Mid-September 2012
LUNCH
SHARING PLATES
MAINS

PODI
THE FOOD ORCHARD

PODI

波蒂美食林是一家全日制餐厅，主营大胆而独特的风味美食。波蒂餐厅是主厨塞德林的作品，秉承了他坚持购买和制作美味而负责任的食品的理念。“波蒂”（Podi）在印度语中代表着一道混合了干调料和香草的菜品，因此餐厅简洁而现代的LOGO也采用了强烈的色彩代表着盛满了各种调料和香草的勺子。大地色系的搭配反映了波蒂餐厅制作美味、自然、有机的食物的追求。

设计机构：Bravo设计公司 创意总监：埃德温·谭 设计师：贾思敏·李 委托方：面包仓库公司

SURABAYA, INDONESIA

SOCIETY

Information Visualisation

社会餐厅 / 印度尼西亚，泗水

信息的可视化

The concept of Society is a restaurant that serves as a meeting platform for various group or individuals with common interests.

Design agency: Thinking*Room Creative directors: Eric Widjaja, Barata Dwiputra Designer: Kenzo Miyake Client: Society

社会餐厅为各种拥有共同兴趣的群体和个人提供了一个交流会面的平台。

设计机构：**Thinking*Room 设计公司** 创意总监：**埃里克·维佳佳、巴拉塔·德维普特拉** 设计师：**三宅贤三** 委托方：**社会餐厅**

桶匠餐厅是一家位于波特兰东南地区中心的酒庄餐厅。餐厅位于一座活动房屋内，独特的建筑结构打造出奇妙的线条和空间，为餐厅的形象设计提供了灵感。精致、现代而充满机械感的室内设计展现了流程与执行的重要性。字体设计同样具有传统的机械效果，而图标则描绘出活动房屋的造型。最终所形成的标识设计能伸能屈，既能被放大装饰整面墙壁，又能栖身于酒瓶之上。

设计机构：Public-Library 设计公司 设计师：拉蒙·科罗纳多、马歇尔·雷克 摄影：蒂娜·阿维拉 委托方：桶匠餐厅

PORTLAND, USA

COOPERS HALL

Traditional Mechanical Effect

桶匠餐厅 / 美国，波特兰

传统的机械效果

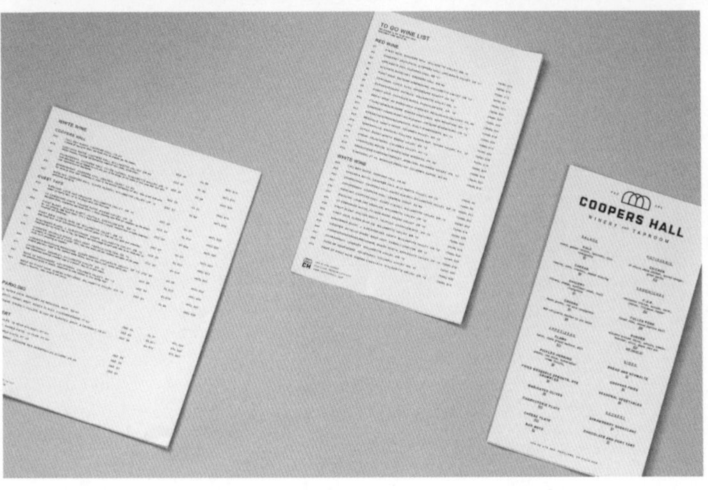

Coopers Hall is a winery and restaurant in Central Southeast Portland. Housed inside a quonset structure, the unique architecture of the building creates unusual lines and spaces that aided in inspiring the identity. A refined, modern mechanical interior shows the importance of process and execution. The letterforms were designed in the tradition of machinery, while the icon represents the quonset form. When paired together it builds a minimal, type driven identity that can evolve and scale to occupy a wall or contract and compliment the curves of a wine growler.

Design agency: Public-Library Designers: Ramón Coronado and Marshall Rake Photography: Dina Avila Client: Coopers Hall

SINGAPORE

OUTPOST 903

A Mix of Vintage and Modernity

前哨 903 酒吧 / 新加坡

复古与现代的混合体

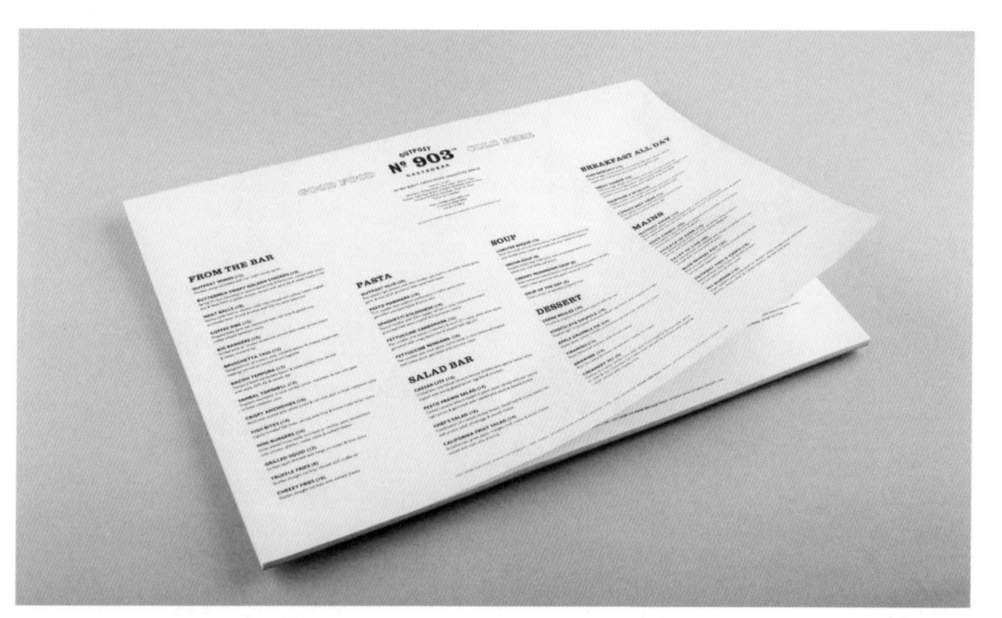

Outpost 903 is a gastrobar located in a 111 years old shophouse that was originally a historic bakery. The interior and branding collaterals were therefore designed to look utilitarian yet contemporary, with old pieces of furniture abandoned by the previous tenant given a new life to keep the original character intact. Raw industrial furniture and fittings were also brought in and customised, with a lofty exposed ceiling completing the look.

Design agency: Bureau Creative directors: Kai Yeo, Edmund Seet, Yasser Suratman Interior consultant: George B.K. Soo, FLIQ Client: Outpost 903

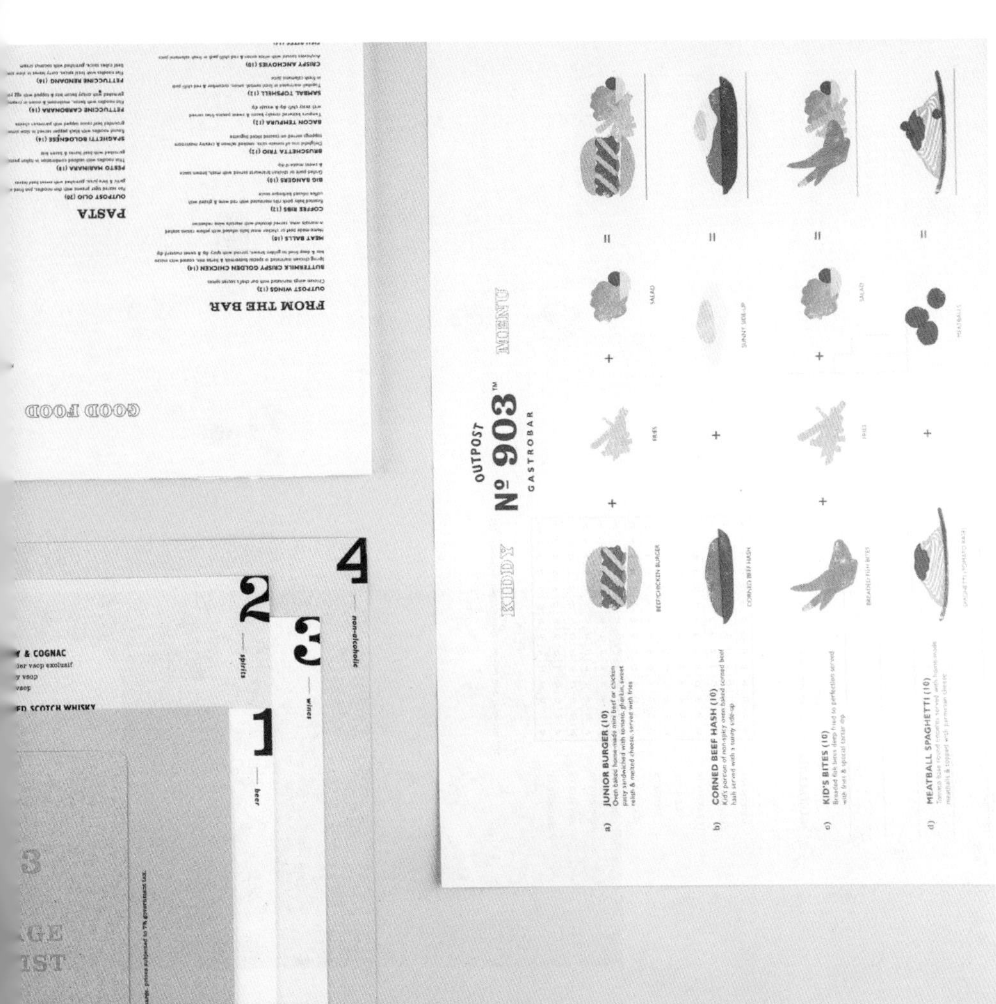

前哨903酒吧是坐落在一座拥有111年历史的旧商店里，商店的前身是一家面包房。酒吧的室内设计和品牌设计实用而现代，设计师为前业主留下的旧家具注入了新的活力，保留了空间的原有风格。同时，设计师还购买和定制了一些工业家具和摆设，配合外露式天花板营造一种复古的氛围。

设计机构：BUREAU ALLS设计公司 创意总监：凯·约、埃德蒙·希特、亚西尔·苏拉特曼 室内设计顾问：乔治·B．K．苏、FLIQ 委托方：前哨903酒吧

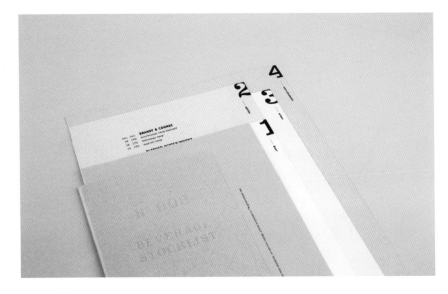

The shape of the Logo refers to the beautiful tiled floor in the restaurant. In combination with art nouveau fancywork, a handwriting and a sans serif type this brings the convivial and nostalgic flair of the great Argentinean coffe-houses into place. The menu is being borne a laser engraved wooden panels. Every information on receipes and letter paper, the designers use a cottonfibre added paper called "Boutique-Wool", is placed there with a stamp. The background of the onepager is a fullscreen gallery which shows the interior of the restaurant in a smooth rotation.

Design agency: ADDA Studio Creative director: Christian Vögtlin Designer: Philipp Vogel Photography: Frederik Laux Client: CAFÉ CHIQUILÍN

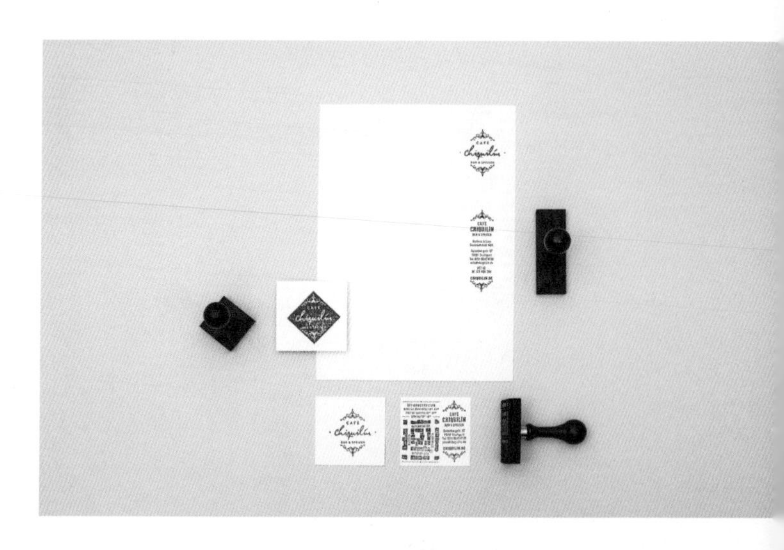

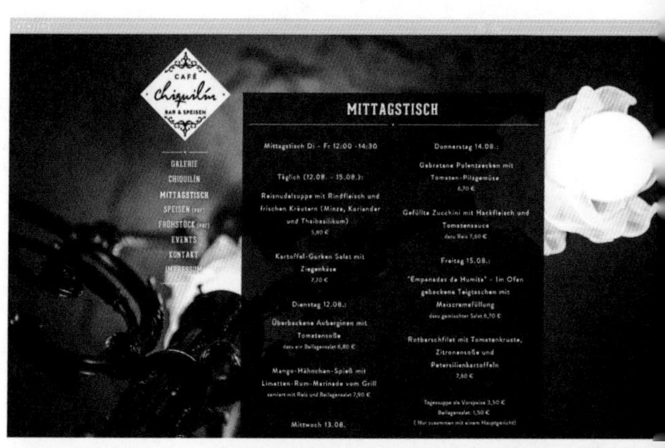

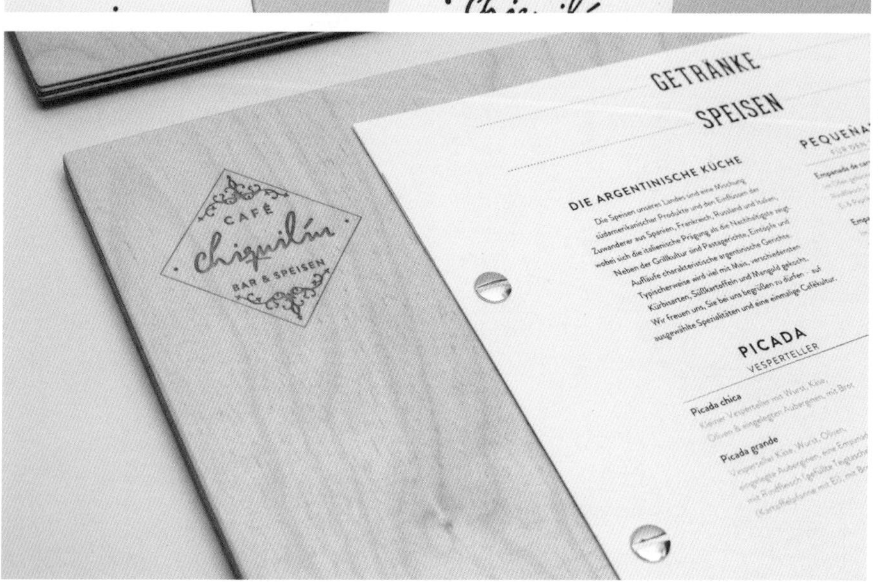

餐厅LOGO的造型参考了漂亮的地砖。新艺术风格的刺绣品与手写无衬线字体为咖啡厅带来了一种欢快、怀旧的气息，就像阿根廷的咖啡馆一样。菜单被贴在经过激光镌刻的木板上，采用的纸张是添加了棉质纤维的"精品毛线纸"，旁边还配有印章。背景的全屏长廊流畅地展示出餐厅的室内设计。

设计机构：ADDA工作室 创意总监：克里斯蒂安·沃特林 设计师：菲利普·沃格尔 摄影：弗雷德里克·劳克斯 委托方：齐克林咖啡厅

STUTTGART, GERMANY

CAFE CHIQUILIN

The Combination of Art Nouveau Fancywork, a Handwriting and a Sans Serif Type to Create a Convivial and Nostalgic Flair

萨尔·库里奥索餐厅 / 德国，斯图加特

新艺术风格刺绣品与手写无衬线字体打造欢快怀旧氛围

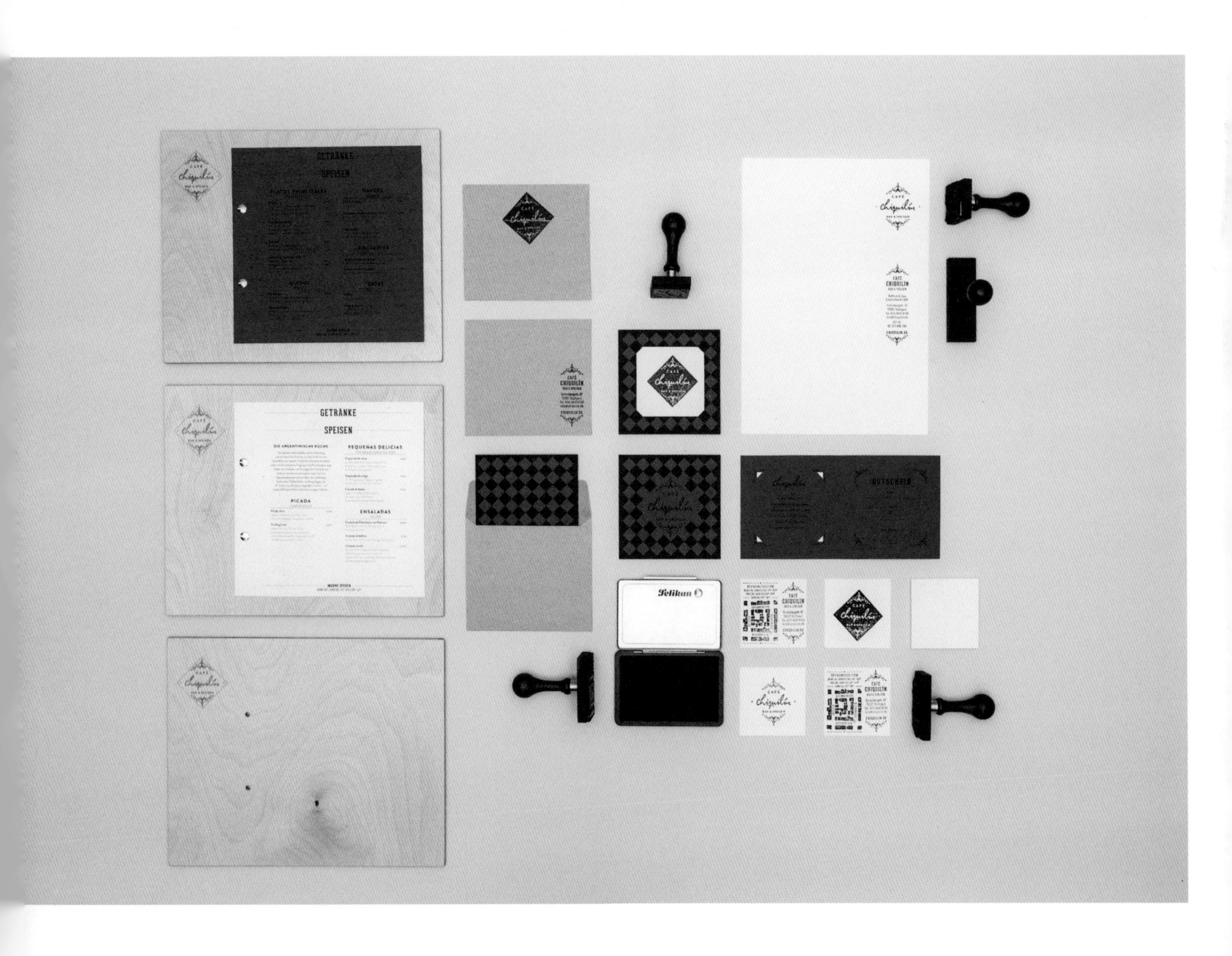

INDEX 索引

& SMITH
062, 184, 186, 240

Abraham Lule & Kuro Strada
022

Acid and Marble
090

ADDA Studio
250

Ageless Galaxy
176, 212

Anagrama
018, 058, 154

Bond Creative Agency
106

Brandon
168

Bravo
034, 038, 064, 086, 088, 262

BUREAU for the Advancement of Lifestyle and Longevity and Success
104

Carpenter Collective
178

Cue
218

Design Ranch
188

Dyer-Smith Frey
068, 070

Eggplant Factory
202

fhancquart.com
190

Foreign Policy Design Group
080, 110

G'DAY Innovation
254

GOMA
096

Haltenbanken
248

Ideogram
142

Indice
216,234,238

Johanna Roussel
072

José Martín Ramírez Carrasco
026

Koyoox
158

La Tortillería
046

Lemongraphic
148

Lo Siento Studio
130, 208

LOKO Design
210

Luda Galchenko
206

antra Branding
42

aria Mileńko
32

asquespacio
50, 118

atteo Morelli, Yurika Omoto,
momi Kuniki, Chinae Takedomi,
suka Miyakoshi,
yoko Aoki
00

emo Productions
50

nd Design
76

odley brand identity
8

gra Nigoević, Filip Pomykalo, Marita
načić
0

New Agency
172, 224

NHOMADA
042

Nychuk Design
198

One&One Design
112, 114, 116

Pixelarte
194

Praktik
200

Public-Library
094, 220, 268

Re-public
246

Rodrigo Aguadé & Manuel Astorga
192

Savvy Studio
028, 040, 092, 122, 126, 134, 138, 182

SNASK
032

Somewhere Else
256, 258

Substance
054, 146

Thinking*Room
266, 267

Tropical
162

Xabier Ogando
196

图书在版编目（CIP）数据

漫食光 ： 餐厅平面与空间设计 / （意）托马斯·拉玛诺斯卡斯编 ；
常文心译. -- 沈阳 ： 辽宁科学技术出版社，2016.1

ISBN 978-7-5381-7396-3

Ⅰ. ①漫… Ⅱ. ①拉… ②常… Ⅲ. ①餐厅－室内装
饰设计 Ⅳ. ①TU238

中国版本图书馆CIP数据核字(2015)第258047号

出版发行：辽宁科学技术出版社
（地址：沈阳市和平区十一纬路29号 邮编：110003）
印 刷 者：利丰雅高印刷（深圳）有限公司
经 销 者：各地新华书店
幅面尺寸：226mm×240mm
印　　张：23
插　　页：4
字　　数：50千字
出版时间：2016年 1 月第 1 版
印刷时间：2016年 1 月第 1 次印刷
责任编辑：关木子
封面设计：关木子
版式设计：关木子
责任校对：周　文
书　　号：ISBN 978-7-5381-7396-3
定　　价：128.00元

联系电话：024-23284360
邮购热线：024-23284502
E-mail：22113829@qq.com
http://www.lnkj.com.cn